AF891861

Esmail Issa

Novel Reactor Design and Method for Atmospheric Pressure Chemical Vapor Deposition of Micro and Nano SiO_{2-x} Films in Photovoltaic Applications

Werkstofftechnik Aktuell

Schriftenreihe aus dem Fakultätsübergreifenden Institut für Werkstofftechnik (IWT) an der TU Ilmenau

Herausgegeben von Univ.-Prof. Dr. rer. nat. Peter Schaaf und Univ.-Prof. Dr.-Ing. Edda Rädlein

Band 26

Die vorliegende Schriftenreihe "Werkstofftechnik Aktuell" berichtet über aktuelle Forschungsergebnisse aus dem Institut für Werkstofftechnik (IWT) der TU Ilmenau. Die ausgewählten Texte spiegeln die breit gefächerten materialwissenschaftlichen und werkstofftechnischen Themen, die am IWT bearbeitet werden, wieder. Für weitere Informationen und Rückfragen können Sie sich gerne an das Institut (www.tu-ilmenau.de/wt) wenden oder das Institut persönlich besuchen. Über Ihre Anregungen, konstruktive Kritik und Ihre Kontaktaufnahme würden wir uns sehr freuen. Das IWT steht wissenschaftlichen Zusammenarbeiten stets aufgeschlossen gegenüber.

Printed and published with the support of the German Academic Exchange Service

This research was prepared at the Fraunhofer Institute for Solar Energy Systems - Freiburg im Breisgau

---

Novel Reactor Design and Method for Atmospheric Pressure Chemical Vapor Deposition of Micro and Nano SiO_{2-x} Films in Photovoltaic Applications

Esmail Issa

Universitätsverlag Ilmenau
2022

Impressum

Bibliografische Information der Deutschen Nationalbibliothek
Die Deutsche Nationalbibliothek verzeichnet diese Publikation in der Deutschen Nationalbibliografie; detaillierte bibliografische Angaben sind im Internet über http://dnb.d-nb.de abrufbar.

Tag der Einreichung:	7. Dezember 2020
1. Gutachterin:	Univ.-Prof. Dr.-Ing. habil. Edda Rädlein (Technische Universität Ilmenau)
2. Gutachter:	Prof. Dr. rer. nat. habil. Thomas Hannappel (Technische Universität Ilmenau)
3. Gutachter:	Prof. Dr. Carsten Ronning (Friedrich-Schiller-Universität Jena)
Tag der Verteidigung:	29. April 2021

Technische Universität Ilmenau/Universitätsbibliothek
Universitätsverlag Ilmenau
Postfach 10 05 65
98684 Ilmenau
https://www.tu-ilmenau.de/universitaetsverlag

ISSN 1868-6532
ISBN 978-3-86360-263-5 (Druckausgabe)
DOI 10.22032/dbt.53692
URN urn:nbn:de:gbv:ilm1-2021000505

Abstract

This research is dedicated to the development of a cost-effective laboratory-scale reactor and method for the atmospheric pressure chemical vapor deposition (APCVD) of SiO_{2-x} films by means of the hydrolysis of $SiCl_4$ at room temperature. The focus for utilization is for photovoltaic (PV) applications, especially crystalline silicon solar cells, where the reduction of fabrication expenses is of high demand. This approach reduces the costs of SiO_{2-x} film deposition as opposed to the well-known chemical vapor deposition techniques. Avoidance of gas-phase reactions, simple reactor design, a wide range of deposition rates, reliable material quality, low safety measures and maintenance times, and stability for the PV applications are the main goals of this work. These goals are achieved.

A novel APCVD reactor made of polycarbonate and thermoplastic materials, is designated to synthesize the SiO_{2-x} films heterogeneously upon the substrate surface with the elimination of the gas-phase reactions. The deposition rate is investigated in accordance with the concentration of the reactant precursors within the media. Thanks to the development of a new evaporation configuration, the reactant concentrations are adjusted precisely before intermixing in an associated injector. The key factors of controlling these concentrations are the temperatures and the flow rates of the reactant gases between the evaporation configuration and the injector.

The SiO_{2-x} films are intended to uniformly coat a substrate area of 6×6 in^2, which is standard for solar cell applications. For this purpose, an APCVD injector design is studied by means of computational fluid dynamics. ANSYS software is used for the relevant simulations. These design investigations are directed towards creating motionless parts, which are sufficient for intermixing the gas precursors homogeneously. The ultimately appropriate design is made of thermoplastic materials using fused deposition modeling technology. The reduction of the gas-phase reactions near the substrate is attempted with a careful design of the injector exhaust and the molar ratio between the reactants. The condensation of reactant precursors and the parasitic oxide deposition upon the APCVD injector's inner surfaces are successfully eliminated without the need for curtain gases common in inline APCVD injectors.

The resultant APCVD SiO_{2-x} films are extensively characterized – in terms of chemical and optical properties – so as to judge their quality and compatibility with PV and other potential applications. The chemical nature and stoichiometry of the films are first investigated with three characterization methods (for verification purposes). The films are shown to be near-stoichiometric and therefore hold the naming SiO_{2-x} instead of SiO_2. The film deposition rate is varied as a function of the operation boundary conditions (flow rate, substrate temperature, and reactant molar-ratio). Variation of the reactant molar-ratio and the substrate temperature (around the room temperature) leads to a wide range of deposition rates and material properties. The hydroxyl contents in the SiO_{2-x} films are detected at various deposition boundary conditions. The calcination for 1 min at relatively low temperatures (< 300 °C) is demonstrated to reasonably remedy the as-deposited films from the hydroxyl groups. The drawbacks of post-deposition calcination at high temperatures (> 500°C) are inspected, as well. A compromise between the reactant molar ratio, substrate temperature, and calcination temperature and duration is investigated to optimize the integrity of the APCVD SiO_{2-x} films, so that they are suitable for various PV applications.

Different approaches, characterized by inexpensive process sequences, are introduced for the fabrication of crystalline silicon solar cells with the assistance of the APCVD SiO_{2-x} films. The APCVD SiO_{2-x} is proposed, after 1 min of calcination, to protect the Si wafer surface from alkaline and acidic solutions, by which a single-side texturing of the Si wafers is obtained. Electroplated metallization schemes of the front side of heterojunction solar cells are tackled with the use of the APCVD SiO_{2-x} as a masking layer where no finger lines are intended. Before the oxide coating, two process sequences for Cu plating are examined. The first process includes seeding with a thin layer of Ag paste to stimulate subsequent plating of the Cu fingers. The second process includes screen printing of a polymer paste in the shape of the later grid electrode. After coating with the APCVD SiO_{2-x} mask, the polymer fingers are striped and electroplating takes place in the resultant openings, directly on the transparent conductive oxide layer. In the last group of applications, the APCVD SiO_{2-x} film performance is investigated for eliminating the parasitic plating on the front side of passivated emitter rear contact solar cells as well as on the rear side of standard multi-crystalline bifacial solar cells.

Zusammenfassung

In dieser Arbeit wurden ein kostengünstiges Verfahren und eine Anlage zur chemischen Gasphasenabscheidung von SiO_{2-x}-Schichten bei Atmosphärendruck (atmospheric pressure chemical vapor deposition, APCVD) im Labormaßstab entwickelt. Dabei kommt die Hydrolyse von $SiCl_4$ bei Raumtemperatur zum Einsatz. Der Anwendungsschwerpunkt für die SiO_{2-x}-Schichten liegt im Bereich Photovoltaik (PV), speziell kristalline Siliziumsolarzellen. Dort ist die Reduzierung der Herstellungskosten von großer Bedeutung. Im Vergleich zu den bekannten Verfahren für die chemische Gasphasenabscheidung senkt der gewählte Ansatz die Kosten für die SiO_{2-x}-Schichtabscheidung deutlich. Hauptziele der Entwicklungsarbeit waren einfaches Reaktordesign, geringe Sicherheitsmaßnahmen und Wartungszeiten, die Vermeidung von Gasphasenreaktionen und Staubbildung, eine für PV-Anwendungen geeignete Schichtqualität sowie die Möglichkeit, die Abscheideraten in einem weiten Bereich zu variieren. Diese Ziele wurden erreicht.

Es wurde ein neuartiger APCVD-Reaktor aus Polycarbonat und thermoplastischen Materialien aufgebaut, mit dem die SiO_{2-x}-Schichten heterogen auf der Substratoberfläche unter Eliminierung von Gasphasenreaktionen synthetisiert werden können. Die Abscheiderate wurde in Abhängigkeit von der Konzentration der Reaktanden im Trägergas untersucht. Dank der Entwicklung geeigneter Verdampferkonfigurationen für die Reaktenden $SiCl_4$ und H_2O können deren Konzentrationen in den inerten Trägergasen vor der Durchmischung in einem Injektor genau eingestellt werden. Die Schlüsselfaktoren für die Kontrolle und Steuerung dieser Konzentrationen sind die Temperaturen und die Volumenströme der Reaktandengase in den Verdampfern und im Injektor.

Das APCVD-Injektordesign wurde mit Hilfe numerischer Strömungsmechanik optimiert. Für die Simulationen wurde die Software ANSYS verwendet. Als Ergebnis der Optimierung können die SiO_{2-x}-Schichten auf einer Substratfläche von 156 × 156 mm^2 gleichmäßig abgeschieden werden. Das ist die derzeitige Standardgröße industriell hergestellter kristalliner Siliziumsolarzellen. Die Design-Studien hatten auch das Ziel, einen Injektor zu entwickeln, der ohne bewegte Teile für eine homogene Durchmischung der Gase sorgt. Das letztendlich geeignete Design wurde aus thermoplastischen Werkstoffen mit Hilfe von 3D-Druck im Schmelzschichtungsverfahren hergestellt. Ferner wurde die Reduzierung der Gasphasenreaktion in der Nähe des Substrats durch Einstellung des Molverhältnisses der Reaktanden und eine geeignete Führung der Injektorabgase erreicht. Die Kondensation von Reaktanden und die parasitäre Oxidabscheidung auf den Innenflächen des APCVD-Injektors wurden erfolgreich vermieden, ohne dass ein bei Inline-APCVD-Injektoren üblicher Gasvorhang erforderlich ist.

Die resultierenden APCVD- SiO_{2-x}-Schichten wurden hinsichtlich ihrer chemischen und optischen Eigenschaften sowie ihrer Zusammensetzung umfassend charakterisiert, um ihre Qualität und Kompatibilität mit PV- und anderen potenziellen Anwendungen zu beurteilen. Dabei zeigte sich, dass die Schichten nahezu stöchiometrisch sind. Deswegen wurde die Bezeichnung SiO_{2-x} anstelle von SiO_2 gewählt.

Die Abscheiderate wurde in Abhängigkeit von den Volumenströmen, der Substrattemperatur und dem Molverhältnis der Reaktanden untersucht. Die Variation der Substrattemperatur nahe der Raumtemperatur und des Molverhältnisses der Reaktanden führt zu einer großen Bandbreite von Abscheideraten und Materialeigenschaften. Die Hydroxylgehalte in den SiO_{2-x}-Schichten wurden bei verschiedenen Abscheidebedingungen bestimmt. Es wurde gefunden, dass die Kalzinierung für 1 min bei relativ niedrigen Temperaturen kleiner 300 °C die Hydroxylgruppen in den abgeschiedenen Filmen deutlich reduziert. Die Nachteile der Kalzinierung bei hohen Temperaturen über 500 °C nach der Schichtabscheidung wurden ebenfalls untersucht. Optimierte Werte für das Molverhältnis der Reaktanden, der Substrattemperatur sowie der Kalzinierungstemperatur und -dauer wurden gefunden, um APCVD-SiO_{2-x}-Schichten ohne mikroskopisch kleine Löcher und Risse zu erhalten, so dass sie für die verschiedenen PV-Anwendungen geeignet sind.

Verschiedene kostengünstige Prozesse für die Herstellung von kristallinen Silizium-Solarzellen unter Verwendung der APCVD SiO_{2-x}-Schichten wurden entwickelt. So konnten nach einseitiger SiO_{2-x}-Beschichtung und einer 1-minütigen Kalzinierung einkristalline Si-Wafer mit alkalischer Ätzlösung einseitig texturiert werden. Eine weitere Anwendung ist die Verwendung von APCVD SiO_{2-x} als Maske für die lokale galvanische Abscheidung des Vorderseiten-Metallkontakts auf Solarzellen mit Heteroübergang. Dabei wurde eine Ag-Paste in Form eines linienförmigen Kontakts mittels Siebdruck auf das transparente leitfähige Oxid (transparent conducting oxide, TCO) der Solarzellen dünn aufgebracht und nach einer ganzflächigen APCVD SiO_{2-x}-Beschichtung der Solarzell-Vorderseite mit Cu galvanisch verstärkt. In einer anderen Prozesssequenz wurde eine Polymerpaste in Form des späteren linienförmigen Metallkontakts mittels Siebdruck auf das Vorderseiten-TCO der Solarzellen aufgebracht und nach der ganzflächigen Beschichtung mit APCVD-SiO_{2-x} mit Lösungsmittel wieder entfernt. In die entstandenen lokalen Öffnungen der SiO_{2-x}-Maske erfolgte die lokale galvanische Metallabscheidung direkt auf dem TCO. In der letzten untersuchten Anwendung wurden APCVD-SiO_{2-x}-Schichten als Schutz vor parasitärer galvanischer Metallabscheidung auf der Vorderseite von einkristallinen *pn*-Solarzellen sowie auf der Rückseite von multikristallinen bifazialen *pn*-Solarzellen untersucht.

Contents

1 Introduction

1.1 Motivation and aim of this research

Chemical vapor deposition (CVD) and atomic layer deposition (ALD) of optical and dielectric/conductive films such as SiO_2, TiO_2, Al_2O_3, Si_3N_4, ZnO, SnO_x etc. are known to possess two main drawbacks – high tool and operation costs and toxic chemical precursors [1–6]. The major tool expense is invested for the vacuum pumps required to obtain low pressure, which is significant for the efficient interaction between the reactant molecules in the deposition chambers. Additionally, the high operation costs are attributed to the need for: i) high thermal energy for the reaction activation, ii) plasma enhancement/electrical frequency generation for the reactant ionizations, iii) ozone production from oxygen, iv) catalysis engagement, v) substrate-surface modification, vi) careful exhaust-gas treatment, and vii) regular tool maintenance and cleaning. The deposition of the mentioned film materials, more specifically, the SiO_2 films utilizes hazardous chemical precursors such as hydrosilane and organosilane products. These products are characterized by high flammability and explosion degrees, leading to high safety measures and hence complex tool architectures. As a result, further costs are added to the tool and operation costs, besides the risks arose on the operating humans.

Deposition of the SiO_2 films on various substrate materials under atmospheric conditions seems to be cost-effective, since no vacuum tools are employed. However, the needed thigh deposition temperatures (around 300 – 600 °C) and the utilized flammable precursors [7,8] are still a serious drawback to be addressed in such a coating industry today. ALD technology utilizes the non-flammable and extremely low-cost chemical precursors H_2O and $SiCl_4$ as an option to prepare stoichiometric SiO_2 films at relatively high temperatures (300 – 500 °C) [9]. The hydrolysis of $SiCl_4$ under atmospheric pressure occurs in a short time of a few milliseconds. This hydrolysis results in intermediate and final reaction product in the gas phase, thus impedes the reaction between the gas species and the substrate surface (known as surface reaction). As a result, oxide particles are generated in the gas medium near the substrate, which fall onto the substrate surface causing an inferior film quality. Figure 1.1 elucidates an inferior silicon oxide film deposited non-uniformly on a shiny etched Si wafer using a linear injector under atmospheric conditions. That is why ALD is a suitable technique when $SiCl_4$ and H_2O reactants are used since it prevents the gas-phase reactions and allows for the reactions between the substrate surface and the two reactants, separately.

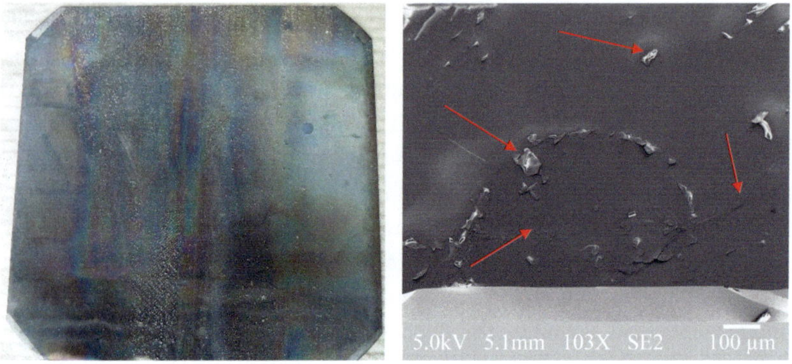

Figure 1.1: Left: Photograph taken for a shiny etched 6-inch large Si wafer after coating with a SiO_x film using a $SiCl_4/H_2O$ injector under atmospheric conditions without preventing the gas-phase impact. Parasitic oxide dust, insufficiently coated area, and inferior thickness uniformity are demonstrated. **Right:** Scanning electron microscopy image taken for a random spot on the top surface of the sample shown in the left photo. It depicts embedded SiO_x particles in the deposited film, flakes repelled from the top surface, and cracks propagated all along.

However, despite its high-quality products, ALD technique is known for its extremely low deposition rates as well as its high investigation. That is because of the alternative introduction-purging of reactant gases and the employment of vacuum pumps needed to generate low pressures of e.g. a few millitorrs. The biggest challenge of this research is to utilize $SiCl_4$ and H_2O vapors for depositing functional SiO_2 films uniformly at room temperature using atmospheric-pressure injectors without generating parasitic oxide particles upon the substrate surface. The development of a reliable system enables reducing the deposition costs of SiO_2 films featuring physical and chemical properties aliening those of ALD and other vacuum-based CVDs.

The aimed deposition system should require low safety measures and simple reactor design without the need for kinetic parts or a high electric power. In order to stand out from other CVDs, the maintenance and cleaning times should be reduced, as well.

The reduction of solar cell manufacturing-costs, besides the increase of their energy conversion efficiency, is of great interest in the current time. For this purpose, several approaches are optimized to shorten the relevant fabrication steps and to continuously make them cheaper. By way of example, replacing the conventional Si wafer texturization with a single-side texturization [10]. Such texturing process requires protecting the rear side of Si wafers with an anti-texturing layer of, e.g. SiO_2 material. Thermally grown by Si surface oxidization, a 10 - 25 nm thick SiO_2 layer is capable of shielding the wafer's rear side from texturing solutions and subsequent phosphorous diffusion. However, growing the protective SiO_2 requires a high thermal budget as well as, in most cases, ultra-low pressure, thus additional costs add up to the overall solar cell fabrication. Another cost reduction example is the replacement of standard grid electrode metallization with cost-effective electroplating schemes [11]. Cu electroplating is proposed as an alternative technology for screen-printing of the expensive Ag pastes. Outstanding conversion efficiency of 26.7 % is achieved by Kaneka cooperation using Cu-electroplated metallization of the bifacial silicon heterojunction solar cells [12]. The new metallization approaches require coating the top surface of the solar cell precursor with dielectric layers for masking the area between the metal finger lines during electroplating. These layers are usually applied utilizing conventional CVD and physical vapor deposition techniques, therefore they necessitate the use of vacuum pumps and high operating temperatures. In this work, new approaches are attempted for the single side texturing of Si wafer and for reducing the costs of electroplated metallization schemes on various kinds of solar cells. For that matter, the newly developed silicon oxide films are to be investigated as a potential candidate for shielding one side of the Si wafers against alkaline/acidic texturing solutions after the shiny etching and cleaning steps. Moreover, the silicon oxide films are to be investigated for selectively insulating the front or/and back sides of different solar cells against the electroplating currents applied for the grid electrode metallization. Consequently, a significant decrease in costs of electroplated metallization is expected to enhance its role as an alternative for screen-printing technology, thus to save costs of Ag consumption on industrial scale level.

1.2 Fundamentals of solar cells

This section reviews the main physics behind solar cells, namely the constructing semiconductors, the recombination of excess carriers, the operation principle, and the parameters of the current-voltage characteristics. Most represented principles are based on the perspectives taken from the textbooks of Würfel et al. [13,14]. This approach states that the solar cell structure is represented by an absorbing semiconductor, wherein the solar radiation is converted to a chemical energy, which is eventually converted to an electrical energy. This absorber features two semi-permeable membranes that transmit electrons and suppress holes at one terminal, whereas at the second terminal, transmit holes and suppress electrons. The selective transport of electrons and holes to the two separate terminals is ascribed to gradients of quasi-Fermi energies. It is shown that the illuminated semiconductor experiences driving the two charge carriers towards both terminals simultaneously. However, the necessary charge carrier selectivity is obtained by differences in the conductivities at the different terminals of the semiconductor. This phenomenon allows e.g. electrons to transport to one contact, whereas they are blocked at the other contact. For detailed reading about solar cell structures and their work principle, it is referred to the textbooks and publications [13 – 21].

1.2.1 Semiconductors and generation of electron-hole pairs

Semiconductors are materials characterized by excitation energy levels interrupted by an energy gap of the width (ε_G). See Figure 1.2. This interruption results in two energy levels, below and above the energy gap. The energy level below the gap is known as the valence band and is virtually completely occupied with electrons before excitation. On the other hand, the energy level above the gap is known as the conduction band and is virtually empty. For an electron in the valence band to be excited by an absorbed photon, the latter must possess an energy $\hbar\omega \geq \varepsilon_G$. Photons with lower energy than ε_G are unable to excite electrons and instead are either transmitted through or reflected by the semiconductor surface similarly to those reflected at metals, wherein the two energy bands are overlapped.

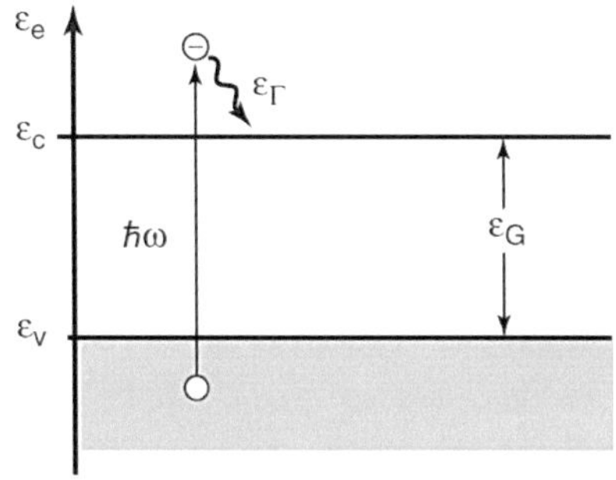

Figure 1.2: A schematic showing excitation of an electron from the valence band into the conduction band of a semiconductor as a result of the absorption of a photon featuring energy $\hbar\omega \geq \varepsilon_G$. From [13].

In order to create an electron-hole pair, a minimum energy of at least ε_G must be introduced to the semiconductor. Seen in Figure 1.2, promotion of an electron from the valence band to the conduction band after absorption of a photon, which transmits its energy ($\hbar\omega = \varepsilon_\gamma$) and momentum ($P_\gamma$) to the semiconductor. This process is realized by the so-called impact ionization, by which an electron or hole possesses a kinetic energy enough to transition out of its bound state in the valance band to an energy state in the conduction band. For that to take place, excitation of the semiconductor network should be provided, e.g. by means of absorbing a photon or inducing a lattice vibration. Photon absorption is controlled by two factors, namely photon energy $\hbar\omega$ and absorption coefficient $\alpha(\hbar\omega)$. The latter is a material dependent property independent of the relevant geometry.

As the electrons are excited into the conduction band, they start losing their gained energy. This loss takes place by gradient generation of phonons, which is insignificant to cross the gap energy back to the valence band. This is due to the fact that electrons gain or lose energy $\hbar\omega$ as a single state equal to or higher than the ε_G. The only possible way for an excited electron to lose energy is to emit a photon or generate a high number of phonons at once, which are less probable to take place. Alternatively, the electron moves in various states of the conduction band for a relatively long time of 10^{-3} s, which is sufficient for the collection and transport of electrons at the terminals and their conversion into electrical current.

1.2.2 Recombination of electrons and holes

The process opposite to electron-hole generation is the recombination, in which the electrons and holes are annihilated. In conformance to the conservation of energy, the annihilation process is accompanied with production of either phonons or photons or of both simultaneously. There are three simultaneous mechanisms of electron-hole recombination, namely radiative band-to-band recombination, Auger recombination, and impurity recombination. The three recombination mechanisms are distinguished by their relevant carrier lifetime (τ). The charge carrier lifetime is the average residence duration of excess carriers in the conduction and valance bands prior to the recombination reaction. This lifetime is defined by the ratio between the charge carrier density (Δn) and the net recombination rate (R) as follows [20]:

$$\tau = \frac{\Delta n}{R} \tag{1.1}$$

The carrier lifetime is affected by all three mechanisms as they occur simultaneously during the recombination process. Therefore, it is evaluated as the average effective lifetime (τ_{eff}) as well as its inverse value is defined as the inverse summation of every individual lifetime (τ_i) [20]:

$$\frac{1}{\tau_{eff}} = \sum_i \frac{1}{\tau_i} \tag{1.2}$$

1.2.2.1 Radiative band-to-band recombination

The radiative recombination process represents the reverse of the photon absorption in the semiconductor. This process includes a spontaneous transition of an electron, temporarily present in the conduction band, to an unoccupied energy state in the valence band. For that, the energy of the electron-hole pair recombination is transformed into a photon which is emitted. The rate of this recombination (R_{rad}) is therefore proportional to the density of free electrons and holes in their relevant states as follows [20]:

$$R_{rad} = B\,(h \cdot e - h_0 \cdot e_0) = B\,(h \cdot e - n_i) \tag{1.3}$$

$$\text{where } h \cdot e = (h_0 + \Delta n) \cdot (e_0 + \Delta n)$$

with B being the radiative recombination coefficient, h and e are the density of the holes and electrons in the thermal equilibrium, respectively, and n_i is the intrinsic carrier density. This recombination is, therefore, controlled by the deviation between the density of the actual carriers and the equilibrium ones. The corresponding carrier lifetime is, thus, defined as follows [20]:

$$\tau_{rad} = \frac{1}{B(h_0 + e_0) + B \cdot \Delta n} \tag{1.4}$$

In case the injection of the minority charge carriers is low, wherein the density of these carriers is below the equilibrium carrier density (usually identified by the concentration of the ionized dopants into the semiconductor), τ_{rad} is, according to equation (1.4), controlled by the quantity of $B\,(h_0 + e_0)$. This translates into $\Delta n << h_0 + e_0 \approx N_{dop}$, and thus, the higher the dopant concentration in the semiconductor, the lower the τ_{rad}. On the other hand, in the case of high injection of the minority charge carriers, by which their density is higher than that of the equilibrium carrier density, τ_{rad} is controlled by the quantity of Δn. This translates into $\Delta n >> h_0 + e_0 \approx N_{dop}$. As the value of B is a material dependent, it is large for direct semiconductors and small for the indirect ones. With crystalline Si being an indirect semiconductor, a phonon or a lattice vibration is involved and, therefore, τ_{rad} is quite large and less likely to be driven by the injection value. Contrary, with GaAs being a direct semiconductor, τ_{rad} possesses a large impact on the charge carrier recombination process.

1.2.2.2 Band-to-band Auger recombination

Auger recombination is an inverse process to the impact ionization, wherein the electron/hole energy is spent to knock an electron from its bond and thus create a new charge carrier. Contrary to the radiative recombination, in Auger one the kinetic energy of the electron-hole pair is not transferred to a photon (no photon emission induced) but rather to an electron or a hole. Therefore, the electron raises to a higher energy state in the conduction band or the hole drops to a lower energy state in the valence band. The gained energy is ultimately spent in the lattice or on collision with phonons. Auger recombination includes interaction between three particles, either one electron with two holes or one hole with two electrons. If the kinetic energy is absorbed by an electron, a set of two electrons and one hole interact during the recombination. The recombination rate ($R_{\mathrm{Aug,e}}$) is, in this case, quantified as follows [13]:

$$R_{Aug,e} = C_e \cdot e^2 \cdot h \tag{1.5}$$

If the kinetic energy is absorbed by a hole, a set of two holes and one electron interact during the recombination. The recombination rate ($R_{\mathrm{Aug,h}}$) is, in this case, quantified as follows [13]:

$$R_{Aug,h} = C_h \cdot h^2 \cdot e \tag{1.6}$$

The net Auger recombination rate (R_{Aug}) is, therefore, the sum of both mentioned rates, which are proportional to the charge carrier densities and are quantified as follows [13]:

$$R_{Aug} = e \cdot h\,(C_e \cdot e + C_h \cdot h) \tag{1.7}$$

where C_e *and* C_h are the Auger coefficients.

Since the interaction between three particles is essential for Auger recombination, the phenomenon is more dominant in semiconductors with high carrier densities. By way of examples, it takes place in highly doped semiconductors or at high injection conditions, and so relies Auger lifetime. ($R_{\mathrm{Aug,e}}$) is, for instance, large for *n*-doped Si, while is ($R_{\mathrm{Aug,h}}$) large for *p*-doped Si.

Auger lifetime for, e.g. *n*-doped Si and low injection conditions ($\tau_{Aug,low}$) is given as follows [20]:

$$\tau_{Aug,low} = \frac{1}{C_e \cdot N_{dop}^2} \tag{1.8}$$

On the other hand, Auger lifetime for high injection conditions ($\tau_{Aug,high}$) is given as follows [20]:

$$\tau_{Aug,high} = \frac{1}{(C_e + C_h)\,\Delta n^2} \quad (1.9)$$

It is deduced that at a low injection of the minority carriers, (τ_{Aug}) is governed by the doping concentration, whereas at a high injection it is governed by the excess carrier density.

Auger recombination is consequently an unavoidable reaction that leads to an energy conversion loss in *p-n* junctions made of *p/n*-doped Si and is, thus, considered the key limitation factor in the Si solar cell efficiency.

1.2.2.3 Defect recombination in the bulk

Bulk-defect recombination is induced by two mechanisms – via impurities that cause defects in various states within the band gap and via lattice defects. Shockley et al. [21] presented a model for quantifying the net recombination in the bulk (R_{SRH}) (measured in $\mathrm{cm}^{-3} \cdot \mathrm{s}^{-1}$), which considers both impurities and lattice defects as follows:

$$R_{SRH} = \frac{h \cdot e - n_i^2}{\tau_{h0}\left(h + h_{1,SRH}\right) + \tau_{e0}\left(e + e_{1,SRH}\right)} \quad (1.10)$$

where τ_{e0} and τ_{h0} are the characteristic lifetimes for the electrons and holes, respectively, which are calculated as follows:

$$\tau_{e0} = \frac{1}{v_{th} \cdot \sigma_{e,SRH} \cdot N_{tr}} \text{ and } \tau_{h0} = \frac{1}{v_{th} \cdot \sigma_{h,SRH} \cdot N_{tr}} \quad (1.11)$$

where v_{th} is the thermal velocity, $\sigma_{e,SRH}$ and $\sigma_{h,SRH}$ are the capture cross section of the electrons and holes respectively, and N_{tr} is the density of bulk defects (or so-called traps).

Also, $e_{1,SRH}$ and $h_{1,SRH}$ are quantities considering the occupation of trap levels as follows:

$$e_{1,SRH} = N_C \exp\frac{(E_{tr} - E_C)}{k_B \cdot T} \text{ and } h_{1,SRH} = N_V \exp\frac{(E_V - E_{tr})}{k_B \cdot T} \quad (1.12)$$

where E_{tr} is energy level of the trap.

E_V and E_C are the energies of the valence and conduction bands, respectively.

N_V and N_C are the state densities in the valence and conduction band, respectively.

k_B is the Boltzmann constant, and T is the device temperature

The lifetime due to bulk recombination (τ_{SRH}), therefore, calculates as follows:

$$\tau_{SRH} = \frac{\tau_{e0}\left(h_0 + h_{1,SRH} + \Delta n\right) + \tau_{h0}(e_0 + e_{1,SRH} + \Delta n)}{h_0 + e_0 + \Delta n} \quad (1.13)$$

After compensating in equation (1.2) the effective lifetime due to bulk recombination calculates as follows:

$$\frac{1}{\tau_{eff,bulk}} = \frac{1}{\tau_{rad}} + \frac{1}{\tau_{Aug}} + \frac{1}{\tau_{SRH}} \quad (1.14)$$

To assess the effect of eliminating either excess charge carriers by induced or diffused junction at the near-surface in comparison with those in the bulk (Δn_{bulk}), effective surface recombination (S_{eff}) is considered. S_{eff} expresses the recombination velocity at the edge of the junction region (also called depletion region or space charge layer) rather than at the top surface.

The ultimately effective lifetime of standard c-Si solar cells, therefore, calculates as follows:

$$\frac{1}{\tau_{eff}} = \frac{1}{\tau_{bulk}} + \frac{1}{\tau_{eff,S}} = \frac{1}{\tau_{bulk}} + \frac{2\,S_{eff}}{W} \quad (1.15)$$

Where $\tau_{eff,s}$ is the effective carrier lifetime at the surface and W is the thickness of the c-Si wafer.

1.2.3 Working principle of silicon solar cells

There are two thermodynamic losses induced in the course of the photovoltaic energy conversion – the loss due to generating electron-hole pairs at the forbidden gap edges and the one induced by entropy generation. See figure 1.4. The entropy generation is governed by three processes, namely the excess carrier generation, their annihilation due to recombination, and the extraction of the remaining excess carriers after recombination. The mean energy of this system is usually defined by the splitting of quasi Fermi-levels of the two distinguishable excess carriers, electrons and holes. The quantity of the Fermi energy splitting is expressed by the so-called implied voltage (iV) as follows [20]:

$$E_{f,e} - E_{f,h} = \frac{iV}{q} = \frac{k_B \cdot T}{q} \ln\left(\frac{(h_0 - \Delta h) + (e_0 - \Delta e)}{n_i^2}\right) \quad (1.16)$$

where $E_{f,e}$ and $E_{f,h}$ are the quasi Fermi-levels of electrons and holes, respectively.

Δe and Δh are the excess electron and hole densities, respectively. q is the elementary charge.

e_0 and h_0 are the equilibrium electron and hole densities, respectively.

Therefore, splitting the $E_{f,e}$ and $E_{f,h}$ explains the difference between excess carrier densities and the equilibrium carrier densities as a consequence of charge carrier generation, their recombination, and extraction of the remaining excess carriers. The higher the splitting of quasi Fermi-levels, the higher the implied voltage resulted, thus a high solar cell output.

Splitting of the quasi Fermi-levels, thus the excess electron and hole densities can be maximized to a further extent by increasing the excess carrier generation, decreasing their recombination, and minimizing the extraction of the remaining excess carriers through external contacts. Consequently, the iV is increased to its maximum. The quantity of implied voltage differs according to the circuit conditions. For open-circuit conditions no excess carriers are extracted from the absorber through the circuit terminals. Therefore, the excess carrier density, thus the implied open-circuit voltage (iV_{OC}), is defined according to the quantification of carrier generation and their recombination, expressed in equation (1.16). However, since no excess carriers are extracted by the open circuit terminals, no electric power is generated, and thus no current flows through. Contrary, under short-circuit conditions, wherein the electrical resistance is nearly zero, the excess carriers are entirely transferred from the electron contact region to the hole contact region through the external circuit. Consequently, no splitting of the quasi Fermi-levels takes place, neither does the voltage generation. For maximum power point conditions the excess carriers are extracted through the external circuit. However, the maximum voltage (V_{mpp}) generated is lower than iV.

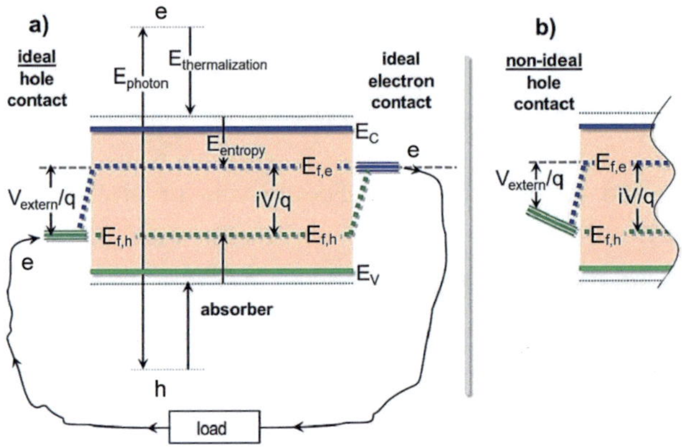

Figure 1.4: a) An energy diagram representing the thermodynamic losses during the photovoltaic process. **a)** ideally selective contacts of the electrons and holes Fermi-levels resulting in external voltage equals to the implied one. **b)** non-ideal hole contact resulting in external voltage below the implied one. After [20].

In ideal contacts shown in Figure 1.4,a) an extraction takes place for the entire quasi Fermi-levels of electrons via its relevant contact and the entire quasi Fermi-levels of the holes via its relevant contact. This ideal energy splitting leads to extracting the iV as a whole from the absorber. However, in real cases the contacts at the absorber terminals are non-ideal and experience energy losses. Shown in Figure 1.4,b), a gradient in the quasi Fermi-level of the hole in its relevant contact occurs. This, subsequently, leads to an extracted voltage value below the iV developed in the absorber. Complete splitting of quasi Fermi-levels of the opposed charge carriers is determined by the selective conductivity of either carrier in its relevant contact region. For instance, if electrons are intended to flow, their local conductivity at their contact region should be significantly high. This eliminates the gradient in the electron quasi Fermi-level and allows the electron current to flow via its terminal contact. Simultaneously, at the same contact region, the local conductivity of the holes (intended to be blocked) should be eliminated to a minimum extent, thus the holes are kept from flowing in the same direction of the electrons. Likewise, the same mechanism applies when it comes to holes flowing via their relevant contact region. The hole local conductivity should be as high as possible whereas the electron conductivity at the same contact region should be eliminated. Consequently, an asymmetry of the local conductivities of electrons and holes is obtained during solar cell operation, hence a strong splitting of quasi Fermi-levels of electrons and holes. As a result, the total current at the electron contact region is a product of flowing electrons in absence of flowing hole in the same direction, and vice versa, the total current at the hole contact region is a product of flowing holes in absence of flowing electrons in the same direction. This is expressed by the following equation [13]:

$$j_{tot} = j_e + j_h = \frac{\sigma_e}{q} \cdot \frac{dE_{f,e}}{dx} + \frac{\sigma_h}{q} \cdot \frac{dE_{f,h}}{dx} \tag{1.17}$$

with j_e and j_h being the current of electrons and holes, respectively.

Such an asymmetry is enhanced by high doping or an induced junction, even during equilibrium (when the solar cell is not operated), hence high excess carrier density in the contact region. However, doping and induced junction of one contact region should provide a high density of one kind of charge carriers (e.g. holes) during operation when additional minority carriers of the other kind (electrons) are generated or injected.

The asymmetry in electron and hole conductivity requires low-injection of minority charge carriers (e.g. electrons) in a contact region. Consequently, the density of the injected/generated minority carriers can be lower than that of the majority carriers (holes in this example) in the same contact region. This, in turn, leads to electron conductivity being lower than the hole conductivity in the hole contact region. Therefore, the electron and hole Fermi-levels are maintained at the two respective terminals. In addition to this maintenance, generating a high voltage at the contact terminals requires a high splitting of the quasi Fermi-levels within the absorber. That is why the contact regions should be as transparent as possible to photons, by which generation of excess charge carriers is extended to its maximum, and subsequently, the recombination of these excess carriers is diminished to its minimum.

Based on what is previously understood, the high extractable voltage from the absorber, thus the high solar cell efficiency, is a trade-off between the asymmetry in carrier conductivity, which is obtained by perfect doping and induced junctions, and the elimination of carrier recombination losses. This is what advanced fabrication of today's solar cells tries to address.

For further reading about the structure and operation principle of solar cells, it is referred to the text books of Würfel *et al.* [13,14] – chapter 6, and the corresponding publication [15].

1.2.4 Current-voltage characteristics

The current-voltage (*I-V*) characteristics are usually determined in order to assess the performance of the solar cell when taken into operation. For that matter, the *I-V* data of the solar cell under illumination are measured and plotted. From the curved *I-V* data the solar cell parameters can be calculated to find: the open circuit voltage (V_{OC}), short circuit current density (j_{SC}), maximum power point (m_{pp}), corresponding current and voltage at the maximum power point (j_{mpp}) and (V_{mpp}), respectively, fill factor (*FF*), and the solar cell conversion-efficiency (η). Figure 1.5 illustrates an *I-V* graph of an illuminated standard mono-crystalline silicon solar cell. It shows the *I-V* curve and the power density (*p*) curve. The non-linear *I-V* characteristics are attributed to the rectifying (non-ohmic) behaviour of the electric current flowing through the *p-n* junction. The V_{OC} and J_{SC} are defined from the intersection of the *I-V* curve with the voltage and current density axes, respectively. The maximum power density of the solar cell m_{pp} corresponds to the obtained J_{mpp} and V_{mpp}. The *I-V* curve measurements are conducted at the standard test conditions:

- The total irradiance on the unit area of solar cells (Φ) is 1000 W/m^2.
- The spectrum of the incident light corresponds to the air mass AM1.5 global (the sum of direct and indirect radiations)
- The temperature of the solar cell is maintained at room temperature.

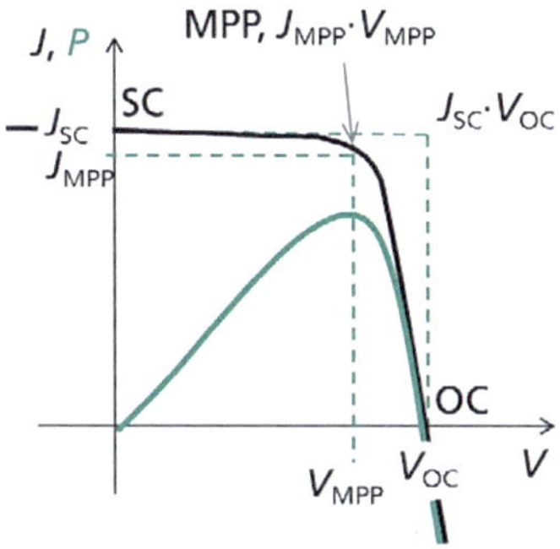

Figure 1.5: *I-V* curve of an illuminated standard mono-crystalline silicon solar cell. From [20].

The *FF* of the solar cell refers to the amount of electrical energy current at the maximum power point (j_{mmp}-V_{mmp}) delivered form the maximum chemical energy current per electron-hole pairs (j_{SC}-V_{OC}). It is calculated from the equation [16]:

$$FF = \frac{J_{mmp} \cdot V_{mmp}}{J_{SC} \cdot V_{OC}} \tag{1.18}$$

FF represents the maximum possible area underneath the *I-V* characteristic filled by the rectangle j_{mmp}-V_{mmp} versus the rectangle J_{SC}-V_{OC}. In practice, *FF* ranges from 0.75 to 0.85.

The maximum possible value of *FF* obtained by a solar cell is called pseudo fill factor (*PFF*). It is determined by measuring the *I-V* characteristic in the absence of external series resistance. The series resistance of a solar cell can be, therefore, calculated with the accounts of both *FF* and *PFF* using the equation:

$$FF = PFF\left(1 - \frac{j_{SC} \,.\, R_S}{V_{OC}}\right) \tag{1.19}$$

The parameters determined form the *I-V* measurements are used to calculate the energy conversion efficiency of the solar cell as follows:

$$\eta = \frac{j_{mmp} \cdot V_{mmp}}{\Phi} = \frac{FF \,.\, j_{SC} \,.\, V_{OC}}{\Phi} \tag{1.20}$$

It prompts the ratio of the electric power generated form the photovoltaic effect to the total luminous received by the solar cell.

1.2.5 The two-diode model

The two-diode model is a simple way for theoretically describing the *I-V* curve of solar cells considered as a large diode. This model is used to investigate the recombination in different regions of the solar cell structure, hence the exponential correlation between voltage and current density at low output voltage up to 0.5 V. The first diode is assumed to express the recombination currents in the emitter and *p*-type base of the cell (direct recombination). The second diode is assumed to express the recombination currents in the depletion region (impurity recombination), or so-called the space charge region. Figure 1.6 illustrates the equivalent circuit diagram for the solar cell [13]. The two-diode model reads:

$$j(V) = j_{01}\left(e^{\left(\frac{q.(V - j\,R_S)}{n_1\,k\,T_{SC}}\right)} - 1\right) + j_{02}\left(e^{\left(\frac{q.(V - j\,R_S)}{n_2\,k\,T_{SC}}\right)} - 1\right) + \frac{V - J\,R_S}{R_P} - j_{SC} \tag{1.21}$$

where:

(j_{01}) and (j_{02}) are the saturation current densities describing the recombination in emitter/base and in the depletion region, respectively.

(n_1) and (n_2) are the ideality factors, which describe the quality of the first and second diodes, respectively. Theoretically, they are $n_1 = 1$ and $n_2 = 2$.

(R_P) is the parallel resistance describing the ohmic losses induced by the current exchange between the electron-hole pairs.

Also, (q) is the elementary charge, (V) is the applied voltage, (k) is the Boltzmann constant, and (T_{sc}) is the solar cell temperature.

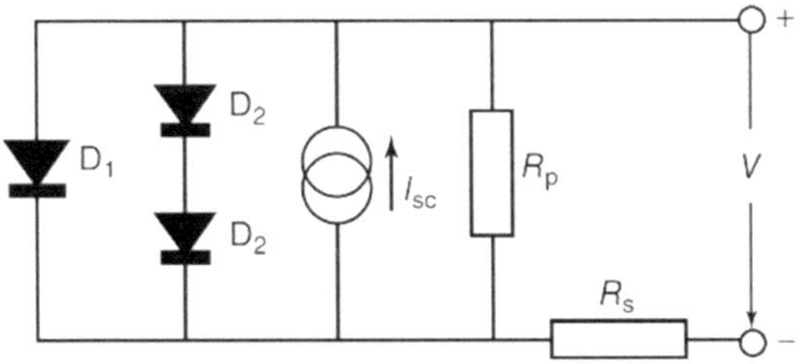

Figure 1.6: Equivalent circuit diagram for a solar cell. It consists of diode D1 representing the direct recombination, diodes D2 representing the impurity recombination, current source I_{SC}, parallel resistance R_P, and series resistance R_S. From [13].

1.3 SiO_2 material preparation-technologies

Mainly, five conventional techniques are employed to chemically prepare SiO_2 layers or particles such as ALD, low pressure chemical vapor deposition (LPCVD), plasma enhanced chemical vapor deposition (PECVD), atmospheric pressure chemical vapor deposition (APCVD), and spray pyrolysis (SP) technique. In this section a brief review of these techniques and their associated reactants are presented. For further reading about SiO_2 structure, properties and preparation technologies, it is referred to the quotations [6,22–26].

ALD of SiO_2 films is performed in ultra-vacuum chambers (10 m Torr) and at deposition temperatures in the range 300 – 527 °C so as to activate the heterogeneous reaction between the reactants and the substrate surface [27]. An initial step is implemented to modify the substrate surface by means of bombarding with Ar beams. The ALD technique implies two half reactions of at least two components for a single deposition cycle. In the first half, one of the reactants is deposited upon the substrate surface, whereas the other reactant is deposited in the second half reaction, at which it interacts with both the firstly deposited reactant and the substrate surface. Subsequently, a nanoscale layer is formed. The deposition cycles are repeated continuously until obtaining the desired film thickness. After every half reaction step, an inert-gas purge of the individual reactant is performed. This causes a wastage of the chemicals since only a fraction of the entire gas volume engages in the deposition. The common chemical precursors used in this technology are H_2O and $SiCl_4$ vapors and the reaction equation is as follows:

$$SiCl_4 + 2\,H_2O \rightarrow SiO_2 + 4\,HCl \tag{1.22}$$

In optimized deposition processes, catalysts such as NH_3 or pyridine are included in the reaction in order to avoid high temperatures necessary for the reactants' dissociation [9,28]. Consequently, more tool investments are required and undesired by-products are present. The ALD techniques are known to yield extremely low film deposition rates which is unsuitable for high throughput processes. This disadvantage makes it inappropriate for applications including polymer materials or requiring low temperatures. This is due to the fact that polymer may be irritated by the vacuum, bombarding beam, catalysts, and the high temperature. The ineffective cost of the SiO_2 ALD is another significant drawback to search for more costly effective deposition technology.

LPCVD of SiO_2 films is performed in a tube furnace under a low pressure of about 0.3 – 0.9 Torr, which is obtained by vacuum pumps. A chemical precursor of diacetoxy-ditertiary-butyl-silane (DADBS) can be utilized at high temperatures ranging from 410 – 600 ºC [29]. The SiO_2 layers can also be deposited by reacting the silane SiH_4 and oxygen O_2 at a temperature of 450 °C and 0.1 Torr based on one of the equations [30]:

$$SiH_4 + O_2 \rightarrow SiO_2 + 2\,H_2 \tag{1.23}$$

$$SiH_4 + 2\,O_2 \rightarrow SiO_2 + 2\,H_2O \tag{1.24}$$

The oxide layers are also LPCVD at a high temperature of 900 ºC by reacting dichlorosilane (SiH_2Cl_2) or silane (SiH_4) with the nitrous oxide (N_2O) gas as follows [31]:

$$SiH_2Cl_2 + N_2O \rightarrow SiO_2 + 2\,N_2 + 2\,HCl \tag{1.25}$$

$$SiH_4 + 2\,N_2O \rightarrow SiO_2 + 2\,N_2 + H_2 \tag{1.26}$$

LPCVD SiO_2 is characterized by high deposition rate of 100 nm/min, conformal step coverage, and free-void layer. Nevertheless, this technology is costly ineffective because of the used vacuum pumps and the high deposition temperatures, which demand complex furnace configuration and high electric power consumption. This induces shortage of the applications, e.g. the ones requiring temperatures lower than 200 °C (when intrinsic a-Si:H is used as passivation for solar cells) or the ones including polymer or metal pastes. This is due to the fact that the diluting solvents essential for subsequent processes would evaporate from the pastes or, even worse, the polymers could burn at the high deposition temperatures. As the LPCVD is molecular diffusion controlled and as the concentration of the reactant precursors decreases during the LPCVD operation, the probability of parasitic dust formation is high and the film deposition rate is inconstant throughout the furnace. The utilization of hydrosilane compounds gives rise to a potential hydrogen-integration in the deposited films, in which structural tensions/stresses are built-up. Since it is flammable and explosive at room temperature once it gets in contact with moisture or air, the hydrosilane compounds causes an instant fire into the exhaust or in the piping prior to the furnace tube. This in turn, increases the safety and maintenance measures adding up to the overall LPCVD expenses.

PECVD is a widely used technique for the deposition of SiO_2 layers. Owing to the presence of plasma, PECVD stands out from other techniques that require high temperatures. The plasma is utilized to dissociate the chemical bonds of the reactants, promoting a completion of the reaction, hence chemically stoichiometric material. The plasma is generated by a high radio frequency (RF) created by voltage application between two electrodes, where one of them is the substrate or the deposition chuck. The reactant precursor molecules charged into the space between the anode and the cathode are ionized and the plasma of the reactants is subsequently generated [31–34]. This is made possible by obtaining either a single high frequency of 13.56 MHz, by which the chuck is earthed, or mixed high and low frequencies of 13.56 MHz on the upper electrode and 100-360 kHz on the chuck, respectively. This is known as double-frequency PECVD. The excitation power needed for such reactors is up to 10 kW. PECVD SiO_2 operates at relatively low temperatures ranging from 25 to 600 °C while the most efficient temperature to deposit pure and stable films is higher than 380 °C. This technique requires an evacuated chamber, wherein the deposition takes place under a pressure of 10^{-7} to 2 Torr. For this purpose, powerful vacuum pumps are needed. Similar to the gas precursors used in the LPCVD technique, hydrosilane compounds are exploited to prepare the PECVD SiO_2.

Based on the above, many drawbacks are present in the PECVD of SiO_2 films. PECVD process is more complex than any other CVD techniques because of the equipment operated to generate different frequencies, by which only the energetic electrons are permitted compared to the ions. The generation of plasma consumes relatively high electric power as it demands many hundreds of watts. However, the most inefficient cost is caused by the vacuum pumps that provide the essential low pressure. Although the temperatures adjusted in this process are relatively low, heating tools are still required, thus more expenses add up to the overall cost. The resulting plasma gas stimulates a potential danger of creating mechanical stress/tension in the deposited layers because of the particle bombardment. The hydrogen contents, may be used in the reactions, cause an integration of Si-H bonds into the films, hence additional built-in tension. As a result, remarkable impurity and porosity concentrations are characterized in the deposited films. The integrated Si-H bonds affect the electrical properties of the SiO_2 films and their absorption of the ultra-violate wavelength (UV), as well. The waste materials through the exhaust and the toxicity and flammability of the hydrosilane compounds increase the safety precautions and hence the overall costs of SiO_2 PECVD technique.

SP deposition of SiO_2 layers is a solution-based technique conducted under the atmospheric pressure either in ambient air [35] or in high-temperature flame (≥ 1600 °C), which is known as the flame spray pyrolysis (FSP) [36]. In the first approach a solution of $SiCl_4$ and pentane-2, 3-dione ($C_5H_7O_2$) is mixed for 3 hours. The resulting solution aerosol is then sprayed onto a substrate heated at a moderated chuck temperature of 350 °C. This process is not continuous, yet is pulsed for 20 sec with a time interval of 30s. The reaction takes place based on the equation:

$$SiCl_4 + 2\ C_5H_7O_2 + 2\ H_2O \rightarrow SiO_2 + 4\ HCl + 2\ H_5C_8O_2 \qquad (1.27)$$

In the second approach TEOS-based solution is sprayed in a premixed methane-air or in n-hexane C_6H_{14} flame.

Numerous disadvantages are induced in the SP deposition technique such as i) low deposition rate due to the pulse manner, ii) non-uniform films due to uncontrollable droplet size, affecting the oxide film quality, iii) large amount of waste solution precursors in the outward stream to the exhaust, iv) relatively high temperature process, v) hazardous, expensive, and flammable materials are used, which requires high safety preparations prior to and in the course of the deposition, vi) oxide dust is expected to be formed as a consequence of the molecular diffusion reaction, which can limit the quality of the deposited films and contaminate the SP apparatus, and vii) essentially regular cleaning/maintenance of the cluttered spray-nozzles and the other parts of the setup.

APCVD is the oldest inexpensive method among CVD technologies. It can be employed for the preparation of SiO_2 films and nanoparticles utilizing various chemical precursors. For instance, SiH_4 and TEOS with oxidants like N_2O, O_3 and O_2 at temperatures range from 125 to 450 °C. APCVD is known to be molecular diffusion controlled, which causes limitations of the relevant applications. This deposition method is overviewed in details in the next chapter.

In additions to the aforementioned deposition techniques, SiO_2 films are prepared by means of i) thermal oxidation of Si wafer surfaces at temperatures of 800 – 1200 °C [37,38], ii) combustion chemical vapor deposition (C-CVD) at temperatures of 1000 – 1300 °C [39,40], and iii) sol-gel deposition (SGD) which exploits liquid precursors at temperatures lower than 100 °C [41,42]. Despite being operated under atmospheric pressure, theses deposition techniques are inadequate for obtaining dense oxide films with reliable stability and properties. Moreover, burning the chemical precursors in C-CVD at such high temperatures induces elevation of the overall costs, besides the probability to develop thermal tensions in the deposited films. In the next chapters, cost-effective method and apparatus are proposed to functions under atmospheric conditions for the preparation of silicon oxide films, which are proven to be stable and reliable for photovoltaic applications.

1.4 Bibliography

[1] B.S.M. Kretzschmar, K. Assim, A. Preuß, A. Heft, M. Korb, M. Pügner, T. Lampke, B. Grünler, H. Lang, Cobalt and manganese carboxylates for metal oxide thin film deposition by applying the atmospheric pressure combustion chemical vapour deposition process, RSC Adv. 8 (2018) 15632–15640. https://doi.org/10.1039/c8ra02288g.

[2] S. Gerullis, L. Pohle, A. Pfuch, O. Beier, B.S.M. Kretzschmar, M. Raugust, E. Rädlein, B. Grünler, A. Schimanski, Structural, electrical and optical properties of SnO_x films deposited by use of atmospheric pressure plasma jet, Thin Solid Films. 649 (2018) 97–105. https://doi.org/10.1016/j.tsf.2018.01.037.

[3] I. Zunke, A. Heft, P. Schäfer, F. Haidu, D. Lehmann, B. Grünler, A. Schimanski, D.R.T. Zahn, Conductive zinc oxide thin film coatings by combustion chemical vapour deposition at atmospheric pressure, Thin Solid Films. 532 (2013) 50–55. https://doi.org/10.1016/j.tsf.2012.11.151.

[4] D. Messerschmidt, A. Hochmuth, E. Rädlein, H. Romanus, W.M. Gnehr, J. Eberhardt, C. Ballif, Anodic degradation of ZnO on soda-lime glass, Sol. Energy Mater. Sol. Cells. 117 (2013) 569–576. https://doi.org/10.1016/j.solmat.2013.07.022.

[5] D. Messerschmidt, Advanced LPCVD ZnO - Challenges in Application for Thin Film Solar Cells and Modules, (2014).

[6] Krishna Seshan, Handbook of Thin Film Deposition Techniques Principles, Methods, Equipment and Applications, Second Editon, Crc Press, New York, 2002.

[7] H. Juárez, M. Pacio, T. Díaz, E. Rosendo, G. Garcia, A. García, F. Mora, G. Escalante, Low temperature deposition: Properties of SiO_2 films from TEOS and ozone by APCVD system, J. Phys. Conf. Ser. 167 (2009). https://doi.org/10.1088/1742-6596/167/1/012020.

[8] J. Kim, S. Hwang, J. Yi, SiO_2 films deposited at low temperature by using APCVD with TEOS/O_3 for TFT applications, J. Korean Phys. Soc. 49 (2006) 1121–1125.

[9] J.D. Ferguson, A.W. Weimer, S.M. George, Atomic layer deposition of SiO_2 films on BN particles using sequential surface reactions, Chem. Mater. 12 (2000) 3472–3480. https://doi.org/10.1021/cm000313t.

[10] S. Gatz, K. Bothe, J. Müller, T. Dullweber, R. Brendel, Analysis of local Al-doped back surface fields for high efficiency screen-printed solar cells, Energy Procedia. 8 (2011) 318–323. https://doi.org/10.1016/j.egypro.2011.06.143.

[11] S.K. Chunduri, M. Schmela, High Efficiency Cell Technologies 2019 - From PERC to Passivated Contacts and HJT, TaiyangNews, (2019).

[12] K. Yamamoto, K. Yoshikawa, H. Uzu, D. Adachi, High-efficiency heterojunction crystalline Si solar cells, in: Jpn. J. Appl. Phys., 2018. https://doi.org/10.7567/JJAP.57.08RB20.

[13] Peter Würfel and Uli Würfel, Physics of Solar Cells: From principles to new concepts. Weinheim: Wiley-Vch Verlag GmbH & Co KgaA, 2005.

[14] Peter Würfel and Uli Würfel. Physics of solar cells: from basic principles to advanced concepts. John Wiley & Sons, 2016.

[15] Uli Würfel, Andres Cuevas, and Peter Würfel, Charge carrier separation in solar cells, IEEE Journal of Photovoltaics. 5 (2014) 461– 469.

[16] Adolf Goetzberger, Joachim Knobloch and Bernhard Voss, Crystalline silicon solar cells, John Wiley & Sons, New York, 1998.

[17] Christian Reichel, Frank Feldmann, Ralph Müller, Robert C. Reedy, Benjamin G. Lee, David L. Young, Paul Stradins, Martin Hermle, and Stefan W. Glunz, Tunnel oxide passivated contacts formed by ion implantation for applications in silicon solar cells, Journal of Applied Physics. 118 (2015) 205701.

[18] Xiaodong Wang and Zhiming M. Wang, High Efficiency Solar Cell. Physics, materials, and devices, Springer Series in Materials Science, Oldenburg. 190, 2014.

[19] Martin. A Green, Solar cells: operating principles, technology, and system applications, Prentice Hall, New Jersy, 1982.

[20] Martin Bivour, Silicon heterojunction solar cells: Analysis and basic understanding, Ph.D. Fraunhofer Verlag, 2017.

[21] W. Shockley and W. T. J. Read, Statistics of the recombinations of holes and electrons, Physical Review, vol. 87, p. 835-42, 1952.

[22] Ralph K. Iler, The chemistry of silica solubility, polymerization, colloid and surface properties, and biochemistry, John Wiley & Sons, 1979.

[23] Roderick AB. Devine, The Physics and Technology of Amorphous SiO_2, Springer Science & Business Media, 2012. https://doi.org/10.5860/choice.41-2870.

[24] C. Robert Helms and Bruce E. Deal, The physics and chemistry of SiO_2 and the Si-SiO_2 interface, New York: Plenum Press, 1988. https://doi.org/10.5749/j.ctt20mvgvz.7.

[25] M.L.H. Anthony C Jones, Chemical vapour deposition: precursors, processes and applications, Royal Society of Chemistry, Cambridge, 2008.

[26] Daniel M. Dobkin and Michael K. Zuraw, Principles of chemical vapor deposition, Springer Science & Business Media, Berlin/Heidelberg, 2003.

[27] J.W. Klaus, A.W. Ott, J.M. Johnson, S.M. George, Atomic layer controlled growth of SiO_2 films using binary reaction sequence chemistry, Appl. Phys. Lett. 70 (1997) 1092–1094. https://doi.org/10.1063/1.118494.

[28] S.M.G. Jason W. Klaus, Ofer Sneh, Growth of SiO_2 at room temperature with the use of catalyzed sequential half-reactions, Science (80). 278 (1997) 1934–1936.

[29] G. Smolinsky, R.E. Dean, LPCVD of silicon oxide films in the temperature range 410 to 600°C from diacetoxyditertiarybutoxysilane, Mater. Lett. 4 (1986) 256–260. https://doi.org/10.1016/0167-577X(86)90018-2.

[30] C. Cobianu, J.B. Rem, J.H. Klootwijk, M.H.H. Weusthof, J. Holleman, P.H. Woerlee, LPCVD SiO_2 Layers Prepared from SiH_4 and O_2 at 450 °C in a Rapid Thermal Processing Reactor, Le J. Phys. IV 5. 5(C5) (1995) C5-1005.

[31] Andrew R. Barron, Chemical vapor deposition of silica thin films, OpenStax-CNX Modul. M24897. 10 (2010) 1–8. https://doi.org/10.3390/cryst10070575.

[32] M. Metzler, R. Patel, Plasma Enhanced Chemical Vapor Deposition (PECVD) of Silicon Dioxide (SiO_2) Using Oxford Instruments System 100 PECVD, 2017. https://doi.org/http://repository.upenn.edu/scn_tooldata/3.

[33] T. V. Herak, T.T. Chau, D.J. Thomson, S.R. Mejia, D.A. Buchanan, K.C. Kao, Low-temperature deposition of silicon dioxide films from electron cyclotron resonant microwave plasmas, J. Appl. Phys. 65 (1989) 2457–2463. https://doi.org/10.1063/1.342815.

[34] M. Zarchi, S. Ahangarani, M.Z. Sanjari, Properties of Silicon Dioxide Film Deposited By PECVD at Low Temperature/Pressure, Metall. Mater. Eng. 20 (2014) 89–96. https://doi.org/10.5937/metmateng1402089m.

[35] M. Esro, O. Kolosov, P.J. Jones, W.I. Milne, G. Adamopoulos, Structural and electrical characterization of SiO_2 gate dielectrics deposited from solutions at moderate temperatures in air, ACS Appl. Mater. Interfaces. 9 (2017) 529–536. https://doi.org/10.1021/acsami.6b11214.

[36] H.D. Jang, H. Chang, Y. Suh, K. Okuyama, Synthesis of SiO_2 nanoparticles from sprayed droplets of tetraethylorthosilicate by the flame spray pyrolysis, Curr. Appl. Phys. 6 (2006) 110–113. https://doi.org/10.1016/j.cap.2006.01.021.

[37] W. Ting, H. Hwang, J. Lee, D.L. Kwong, Growth kinetics of ultrathin SiO_2 films fabricated by rapid thermal oxidation of Si substrates in N_2O, J. Appl. Phys. 70 (1991) 1072–1074. https://doi.org/10.1063/1.349701.

[38] H.Z. Massoud, J.D. Plummer, E.A. Irene, Thermal Oxidation of Silicon in Dry Oxygen: Growth-Rate Enhancement in the Thin Regime: II . Physical Mechanisms, J. Electrochem. Soc. 132 (1985) 2693–2700. https://doi.org/10.1149/1.2113649.

[39] P. Rüffer, A. Heft, R. Linke, T. Struppert, B. Grünler, Characterisation of thin SiO_x-layers on float glass deposited by Combustion Chemical Vapour Deposition (C-CVD), Surf. Coatings Technol. 232 (2013) 582–586. https://doi.org/10.1016/j.surfcoat.2013.06.031.

[40] T. Struppert, A. Heft, B. Grünler, Thin functional films by combustion chemical vapour deposition (C-CVD), Thin Solid Films. 520 (2012) 4106–4109. https://doi.org/10.1016/j.tsf.2011.06.048.

[41] M. Vishwas, K.N. Rao, K.V.A. Gowda, R.P.S. Chakradhar, Optical, electrical and dielectric properties of TiO_2-SiO_2 films prepared by a cost effective sol-gel process, Spectrochim. Acta - Part A Mol. Biomol. Spectrosc. 83 (2011) 614–617. https://doi.org/10.1016/j.saa.2011.08.009.

[42] A. Matsuda, Y. Kotani, T. Kogure, M. Tatsumisago, T. Minami, Transparent anatase nanocomposite films by the sol-gel process at low temperatures, J. Am. Ceram. Soc. 83 (2000) 229–31. https://doi.org/10.1111/j.1151-2916.2000.tb01178.x.

2 Overview of the APCVD SiO_2 technologies

2.1 Introduction

Chemical vapor deposition (CVD) under atmospheric pressure (APCVD) is one of the most common gas-phase chemical processes to date, and the primarily predominant technique to grow SiO_2 films [1]. Chronologically, APCVD is the first CVD technology developed in the (the 1880s) to generate carbon fibers initiating from a carbon filament through thermal activation in an organic solution [2]. Since APCVD systems operate under atmospheric pressure (independent of vacuum systems) and at temperatures below 1000 ºC (practically in the range of 25 – 500 ºC for the majority of them), they are classified among the lowest cost-investment as well as simplest CVD technologies [3–5]. APCVD is characterized by high throughput, high deposition rates up to 500 nm/min, and satisfactory film uniformity and step-coverage conformity [1,3]. In contrast to thin solid films prepared at high-temperatures, ranging from 700 to 1300 ºC such as in low pressure CVD (LPCVD) and thermal oxidation, no thermal stress or deflection is induced in low-temperature APCVD films/substrates [6,7]. Thanks to the ability of deposition at low temperatures, this advantage is elucidated by the absence of residual stress when cooling the deposited substrates to room temperature.

The extensive utilities of the APCVD system, in industry and researches, make it of great interest, as do various materials deposited this way e.g. metallic, organic or inorganic material-based films. The deposited matters can be sorted to oxide and non-oxide films, as well. By way of examples on non-oxide materials, APCVD technology is employed to coat cutting inserts, e.g. cemented carbide-based tools, with hard-wear and anticorrosion layers of TiC, TiN, Cr_7C_3, FeB, Fe_2B SiC or Si_3N_4 (thickness can be a few micrometers) [8]. Since this coating is performed upon thick substrates of metallic-constructed objects and since their edges are shaped/sharpened, high depositing temperatures of 800-1000 ºC are not detrimental to their structure. It is also possible to exploit APCVD technology for growing epitaxial Si and Si/Ge films for semiconductor applications [9]. The deposition temperature can be brought down to 550 °C and the film thickness of up to 3 μm can be realized. Since it is conducted under ambient conditions, oxidation at the Si-growth interface and carbon contamination are impeded by subjecting to the hydrogen environment, despite using ultra-clean chambers. In addition to the grown epitaxial Si, it is possible to dope it, using the same APCVD system, with phosphorous and boron to form *n*-type and *p*-type doped layers, respectively [9]. Si thin layers of 35 – 178 nm are proven to be APCVD formed upon flat glass ribbons during manufacturing, while still being at high temperatures (below the melting point) [10]. Such functional layers are utilized for preventing the infrared (IR) spectrum, preferably of the wavelength $\lambda = 10$ μm, from being transmitted through glass windows. This layer, embedded in a stack of double silver layers and an anti-reflection coating, permits the visible and near-IR spectrum through the glass while reflecting the mid-IR spectra back, hence passive house heating. Si_3N_4 compounds, to be used as a passivation layer for glass and integrated optical devices, are often synthesized by APCVD utilizing a mixture of SiH_4 and NH_3 [11,12].

APCVD technology is also significantly employed for the preparation of oxide thin films to be exploited in numerous disciplines. These films are mainly divided into transparent conducting oxides (TCOs) and dielectric oxides. APCVD TCO films have been shown to be made of different materials; so far instances: zinc oxide (ZnO) [13], tin oxide (SnO_2) [14], fluorine-doped tin oxide (FTO) (SnO_2:F), tin-doped indium oxide (ITO), tin/fluorine-doped indium oxide (ITO:F) ($In_2O_3 \cdot SnO_2$:F) [15], and fluorine-doped zinc oxide (FZO) (ZnO:F) [16]. Such TCO films have been shown to deliver high performance when designated for glass-based thermoelectric (TE) devices [16], liquid crystalline displays (LCDs) [17], and Si/glass-based solar cells [18]. FTO films have been proven to efficiently form pyramidal texture, directly in the course of modified APCVD upon glass and Si substrates [19–21]. Microcrystalline or amorphous Si tandem solar cells, textured this way, demonstrate high light trapping, hence high cell conversion efficiency. Dielectric oxide films can be effectively prepared with APCVD systems for numerous functions, as well. For instance, AlO_x films, deposited this way, can function as a surface passivation layer for silicon solar cells [22]. APCVD TiO_x films have been investigated to play an adequate anti-reflection coating (ARC) role for Si solar cells [23,24]. SiO_x/SiO_2 featuring wide-range physical properties is the most predominant material when it comes to APCVD utilization. By way of examples, but not limitation, SiO_x/SiO_2 was shown to compose, under atmospheric pressure, in the form of nanoparticles that act as e.g. support for metal catalysts, an agent in thermal insulators, an H_2O absorber in humidity sensor, etc. [25]. Fehlner, *et. al* [26] proved the compatible employment of APCVD SiO_2 films in LCD productions. Other authors have demonstrated the advantageous utility of APCVD SiO_2 films in fiber optics and optoelectronic integrated circuits, as well as, in ARC structuring upon glass panels and solar cells [27–30]. Thanks to their high electrical resistivity (ρ = 1021 $\mu\Omega$.cm for the thermal kind [1]) and excellent planarization, APCVD SiO_2 films have been employed in semiconductors and solid-state devices as e.g. i) a gap-filling insulator for inter/pre-metal dielectric films ii) a borophosphosilicate glass (BPSG) interlayer dielectric film iii) and a passivation layer for poly-Si thin-film transistors (TFTs) [31–33]. Both surface passivation of Si precursors and boron or phosphorus-doped SiO_2 layers generated employing APCVD is of increasingly great interest for PV applications [34,35]. This transformation from vacuum-based CVD to APCVD tools is a promising approach for the preparation of such functional SiO_2 films, particularly for reducing costs of solar cell manufacturing, which is our main area of focus for the current time. However, searching new chemicals, minimizing the required safety measures, and simplifying conventional SiO_2 APCVD configurations, are of continuously growing demand. Ultimately, further reduction of APCVD overall costs with maintaining sufficient film quality is to be obtained for solar cell applications.

The aforementioned APCVD SiO_2 films, particularly the ones utilized for solar cell applications, are mainly based on either hydrosilane compounds such as SiH_4 or organic sources such as tetra-ethyl-ortho-silicate, or known as tetra-ethoxy-silane (TEOS) ($Si(OC_2H_5)_4$). These chemicals produce SiO_2 in the presence of ozone (O_3) or O_2 at elevated temperatures up to 500 ºC [34,35]. Although named low-temperature process, films deposited utilizing these chemicals still necessitate high thermal budget and, even worse, sophisticated safety measures since the chemicals used are extremely flammable and hazardous. Additionally, their relevant reactors are complex to design and maintain, particularly when the multiple-injector system is adopted, and they require periodical cleaning contributing to the ultimate process costs [1,5]. Due to the simultaneous exhaustion of the reaction by-products and possible particles, to impede their embedment upon the film, high precursor wastage is expected [1]. The prerequisite deposition temperature adversely affects the polymer-processed or metal paste-coated substrates, which are essential for recently developed PV applications e.g. plating metallization of solar cells [36,37]. Furthermore, oxide particle/dust generation, hence, film contamination is one of the major drawbacks and require additional measures to overcome [38,39].

2.2 APCVD SiO_2 precursors and methods

This section summarizes the most common deposition methods for the formation of APCVD SiO_2 films. There are currently numerous atmospheric reaction dynamics for the preparation of SiO_2 films with different chemical precursors, reaction parameters, and reactor arrangements, which prompt various product properties.

Silane gas, due to its pyrophoric property, abruptly reacts with oxygen forming SiO_2 [1,4]. In order to minimize the gas-phase reaction implications, the gas precursors are therefore injected separately to the reaction zone of the apparatus. The equation characterizing this reaction is as follows:

$$SiH_4 + O_2 \rightarrow SiO_2 + 2\ H_2 \uparrow \qquad (2.1)$$

SiH_4 is widely utilized to form amorphous SiO_2 films to be used as a dielectric interlayer e.g. in the fabrication of dynamic random-access memory (DRAM) and logic devices when low moisture content is induced in the deposited film. However, when abundant O_2 is dosed to the reactor, H_2O content is present in the SiO_2 films as follows:

$$SiH_4 + 2\ O_2 \rightarrow SiO_2 + 2\ H_2O \uparrow \qquad (2.2)$$

The reactions in equations (2.1) and (2.2) are exothermic while the substrate temperature required to initiate the reactions is in the range of 300-

500 °C which results in deposition rates of 200-300 nm /min. The higher the substrate temperature set, the higher the resulting deposition rate. The higher substrate temperature also results in lower thickness uniformity and step coverage conformity induced when filling high-aspect-ratio gaps. To address this trade-off, particularly appearing at film thickness of 0.5 to 1 μm, the deposition temperature is usually set at 400 °C and the film is subsequently subjected to a calcination treatment at elevated temperature of 700 °C. This process permits outgassing the H_2 and OH impurities and thus densifying the deposited SiO_2 films. For instance, H_2 content (in the form Si-H) of such films is in the range of 1-2 %. Deposition rate and thickness uniformity of the silane-based SiO_2 films are sensitive to other parameters such as gas reactants' flow and exhaust flow. The deposition rate is proportional to the flow rate of both SiH_4 and O_2 gases whereas it is less sensitive to the exhaust flow. In contrast, the film thickness uniformity is reversely proportional to the reactant flow rates and is quite sensitive to the exhaust flow, e.g. non-uniform flow of the gases over the entire substrate surface is induced at high-speed exhaust. At balanced reactant flow-rates and exhaust flow the thickness uniformity is ±3 % at a thickness range of 100-1000 nm. Besides its limited filling of high aspect-ratio gaps (of > 1:1), SiO_x particles are incorporated into the deposited film as a consequence of the gas-phase reaction characterizing the APCVD technique. Rough-surface films are therefore obtained and particle contaminants are accumulated upon the reactor and exhaust inner walls. Films prepared this way feature a refractive index of 1.45 and a dielectric strength of 6-7 MV/cm.

Nitrous oxide is another oxidizing reagent for silane -- when reacting together SiO_2 films can be deposited under atmospheric pressure at substrate temperatures ranging from 495 to 690 °C, higher than that when O_2 is used [1,40]. N_2O was found to be a good alternative to avoid dust formation resulted by the highly reactive O_2. The relevant reaction takes place according to the following equation:

$$SiH_4 + 2\,N_2O \rightarrow SiO_2 + 2\,H_2 + 2\,N_2\uparrow \qquad (2.3)$$

Figure 2.1 shows the SiO_2 deposition rate as a function of substrate temperature for silane chemistry with different oxidants [1]. While maintaining the reactor walls relatively cold it is proven that the higher the reaction temperature, the higher the deposition rate for the reactant mixture of SiH_4 and N_2O. When ozone is used as an oxidant the temperature required for dissociating molecules and thus initiating the reaction is quite low in comparison with the other oxidants and the increase in the SiO_2 deposition rate is linear. Despite the high amount of thermal energy provided to disassociate SiH_4/N_2O molecules, an H_2 impurity level higher than that of SiH_4/O_2 is present in the SiO_2 film.

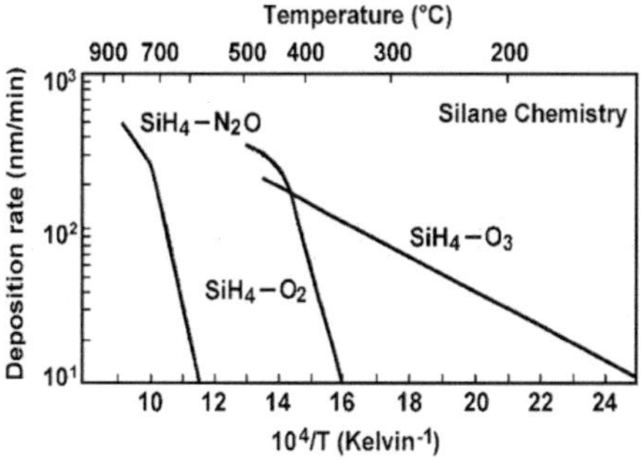

Figure 2.1: A graph showing the deposition rate of SiO_2 films as a function of the substrate temperature when SiH_4 chemistry is used with three different oxidants. (From [1])

TEOS $Si(OC_2H_5)$, apart from being inexpensive, is a way more safer precursor for the formation of SiO_2 in comparison with SiH_4 chemistry [4]. It is a stable, non-pyrophoric, and non-corrosive liquid and its toxicity and flammability are the same as those of alcohols. The first way to produce SiO_2 films under ambient pressure is to react with O_2 according to the following equation [4]:

$$Si(OC_2H_5)_4 + 12\ O_2 \rightarrow SiO_2 + 10\ H_2O + 8\ CO_2 \qquad (2.4)$$

However, because the reactivity of O_2 is relatively low at pressure of 1 atm, decomposition of TEOS is extremely slow and therefore requires deposition temperatures in excess of 750 °C. TEOS is a liquid at room temperature and its vapor pressure is sensitive to the temperature. Consequently, it is dosed through an insulated evaporator with precisely controlled flow rates. Deposition rate of $TEOS/O_2$ films increases linearly with increasing the substrate temperature (seen in Figure 2.2). The SiO_2 films deposited this way feature an excellent step coverage conformity and gap filing. This is due to the fact that a little gas-phase reaction takes place prior to the substrate surface, which in turn results in smoother film surface than deposited by SiH_4 chemistry. Thanks to high deposition temperatures, densification of the films occurs simultaneously during the process and thereby the H_2/OH contents are negligible. Nevertheless, limited film utilization is disadvantageously induced e.g. when films are to be deposited on metal or other temperature-sensitive substances.

As an alternative oxidant for TEOS, ozone has been employed so as to reduce the thermal budget required for increasing O_2 reactivity at atmospheric pressure [1,4]. Since O_3 is a triatomic molecule its unstable characteristic allows for dissociation of TEOS more efficiently than O_2 at lower temperatures of 400 °C. An example on the reaction between ozone and TEOS is the attack of the alkyl group (C_nH_{2n-1}) with e.g. silanol (Si-OH) as represented in the following equation:

$$\text{Si-O-CH}_2\text{CH}_3 + \text{O}_3 \rightarrow \text{Si-O-H} + \text{CH}_3\text{CHO} + \text{O}_2 \qquad (2.5)$$

As such SiO_2 films are used in the production of DRAMs and poly-silicon TFTs due to in-situ planarization as well as dielectric and passivation properties [4,33]. The deposition rate of these films increases directly proportional to the increase in the deposition temperatures. Figure 2.2 depicts the relationship between deposition temperature and resultant deposition rate of TEOS/O_3 system is obtained is 380 ° C, which is in the range of 100-200 nm/min, whereas increasing the temperature beyond 400 °C causes decrease in the deposition rate. This is due to the fact that ozone concentration in the TEOS/O_3 system decreases at temperatures higher than 400 °C.

The deposition rate of the SiO_2 films, on the other hand, is proportional to TEOS/O_3 flow rates as well as to the exhaust flow. The film thickness-uniformity decreases at elevated deposition temperatures above 700 °C and is extremely sensitive to the exhaust flow. Nevertheless, at precisely adjusted deposition temperature (400 °C) the variation of thickness uniformity is ±3 % for 100-1000 nm thick films. The sufficient surface mobility of the mentioned reactant species allows for excellent step coverage conformity and gap filling.

A disadvantage to the TEOS/O_3 system is the high presence of H_2 impurity in the deposited films (9 to 12 at %) in the form of Si-OH and H_2O groups. This is due to both in-situ formed silanol groups in TEOS and relatively lower deposition temperatures (compared to those set for TEOS/O_2 system), which provides insufficient energy to outgas the excessive species. However, this drawback can be addressed by densification at elevated temperatures of 500-700 °C for 1 min [4], [33]. The refractive index and the dielectric strength of TEOS/O_3 films are measured to be 1.44 and 5-6 MV/cm, respectively.

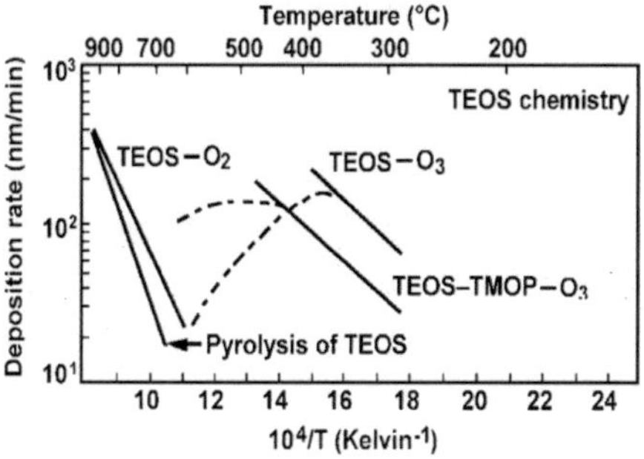

Figure 2.2: A graph showing the deposition rate of SiO_2 films as a function of the substrate temperature when TEOS chemistry is used in different systems. Note that TMOP stands for trimethylphosphate. (From [1])

Other reactant precursors used to prepare SiO_2 films by APCVD technology are $SiCl_4$ and O_2. The oxidization of $SiCl_4$ requires high temperatures in the range of 1200-1400 °C [41] following the reaction:

$$SiCl_4 + O_2 \rightarrow SiO_2 + 2\ Cl_2 \tag{2.6}$$

However, the relevant deposition rate increases gradually with the substrate temperature to reach the maximum of about 1nm/min at temperatures between 1227 °C and 1327 °C, higher than which the rate decreases. This effect is because at temperatures lower than this range the deposition is dominated by SiO_2 monomers formation, by which high quality films (smooth surface, conformality and purity) are obtained. On the other hand, at higher temperatures SiO_2 particle formation is dominated, particularly closer to the reactant introduction-outlets leading to thinner films and, even worse, particle contamination upon the deposited surface. Similar to the other deposition methods, film thickness uniformity is governed by the precise control of uniform temperature profile all over the substrate area.

2.3 Drawbacks of the conventional APCVD methods of SiO_2 films

There are often encountered impediments to the conventional SiO_x APCVD systems and methods, which are summarized as follows:

1- Demand of thermal energy necessitated (at atmospheric pressure) for decomposing the reactant molecules (TEOS, SiH_4, N_2O and O_2) and increasing their mobility/diffusion, thus activating the resultant chemical reaction [1]. For instance, the activation energy required for silane-based reactions is in the range of 10 eV.

2- High-temperature depositions prevent processing heat-sensitive materials like polymers, metal coatings and interlayers, passivation layers such as hydrogenated-amorphous Si e.g. used in SHJ solar cells, and different low-temperature evaporated/melted substances [4][42].

3- As for relatively low-temperature formation (in the range of 250-400 °C) of SiO_2 films or micro/nano-sized particles, a calcination temperature up to 700 ºC and a treatment time up to several hours are essentially prompted. This is to eliminate the parasitic impurities, e.g. H_2O molecules and OH groups, in the final product and to improve its structural density and electrical properties [25][33].

4- Hazardous and explosive substances are employed such as N_2O and SiH_4, thus high safety measures and APCVD system complexity are essential [1,5].

5- A complex post-process handling of the reaction by-products and non-reacted precursors is required to prevent fire/explosion in the exhaust and toxification of the environment. Various hazards combined with venting and treating the SiH_4 gas, as well as, solutions to safe disposal thereof are reviewed by VanOmmeren *et al.* [43].

6- Dust formation common in conventional APCVD technology which is addressed by inefficient reactor arrangements e.g. i) temperature variation between the substrate and reactor walls to prevent parasitic deposition in the reactor interior, ii) curtain flows between and around the reactant gases to restrict the reaction upon the substrate surface, iii) or high flow velocities of the reactant gases allowing for short-term residence in the reaction zone [1,5].

7- Expected high wastage of reactant precursors released through the exhaust on account of the measures taken to eliminate dust formation and particle contamination of the deposited films.

8- Due to high temperatures used, most of the well-known APCVD tools are manufactured of metal components [1,5]. This gives rise to the risk of metal particle-contamination and hence degradation when it comes to processing Si-based devices such as solar cells, microchips, transistors, etc.

Based on the above, effective costs of the overall SiO_2 deposition process are increased because of the complexity of APCVD tools, high wastage of the expensive reactant precursors, high thermal budget needed for activating the reaction and densifying the final products, and short-interval maintenance and cleaning of the related chambers.

2.4 Bibliography

[1] Krishna Seshan, Handbook of Thin-Film Deposition Processes and Techniques, William Andrew, New York, 2001.

[2] W.E. Sawyer, O.F.N.E.W. York, A. Man, O.F. Broorlyn, A. To, E. Dynamic, Carbon for electric lights, United States Patent Office, 2009. https://doi.org/10.1111/j.1559-3584.1927.tb04229.x.

[3] N. Grant, K. Mcintosh, A review on low temperature chemically formed silicon dioxide for solar cell applications, in: 48th AuSES Annu. Conf., Canberra, ACT, Australia, 2010: pp. 1–10.

[4] D.M. Dobkin, M.K. Zuraw, Principles of Chemical Vapor Deposition, 2003. https://doi.org/10.1007/978-94-017-0369-7.

[5] M. Scharfstein, Gaurf, Chemical Vapour Deposition (CVD), Advances, Technology, and Applications, 2013. https://doi.org/10.1017/CBO9781107415324.004.

[6] S.C.H. Lin, I. Pugacz-Muraszkiewicz, Local stress measurement in thin thermal SiO_2 films on Si substrates, J. Appl. Phys. 43 (1972) 119–125. https://doi.org/10.1063/1.1660794.

[7] C. Malhaire, M. Le Berre, D. Febvre, D. Barbier, P. Pinard, Effect of clamping conditions and built-in stresses on the thermopneumatic deflection of SiO_2/Si membranes with various geometries, Sensors Actuators, A Phys. 74 (1999) 174–177. https://doi.org/10.1016/S0924-4247(98)00312-4.

[8] H.E. Hintermann, Tribological and protective coatings by chemical vapour deposition, Thin Solid Films. 84 (1981) 215–243. https://doi.org/10.1016/0040-6090(81)90021-3.

[9] T.O. Sedgwick, P.D. Agnello, Atmospheric pressure chemical vapor deposition of Si and SiGe at low temperatures, J. Vac. Sci. Technol. A Vacuum, Surfaces, Film. 10 (1992) 1913–1919. https://doi.org/10.1116/1.578115.

[10] P. Chesworth, Coating glass, United States Patent Office, 1984.

[11] T. Otani, M. Hirata, High rate deposition of silicon nitride films by APCVD, Thin Solid Films. 442 (2003) 44–47. https://doi.org/10.1016/S0040-6090(03)00938-6.

[12] O.P. Agnihotri, S.C. Jain, J. Poortmans, J. Szlufcik, G. Beaucarne, J. Nijs, R. Mertens, Advances in low temperature processing of silicon nitride based dielectrics and their applications in surface passivation and integrated optical devices, Semicond. Sci. Technol. 15 (2000) R29–R40. https://doi.org/10.1088/0268-1242/15/7/201.

[13] J. Van Deelen, A. Illiberi, B. Kniknie, E.H.A. Beckers, P.J.P.M. Simons, A. Lankhorst, Atmospheric pressure chemical vapor deposition of ZnO: Process modeling and experiments, Thin Solid Films. 555 (2014) 163–168. https://doi.org/10.1016/j.tsf.2013.08.009.

[14] K. Lagha-Menouer, M.S. Belkaid, M. Pasquinelli, Electrical, optical and structural properties of tin oxide thin films deposited by APCVD, Adv. Mater. Res. 651 (2013) 38–41. https://doi.org/10.4028/www.scientific.net/AMR.651.38.

[15] B. Mayer, Highly conductive and transparent films of tin and fluorine doped indium oxide produced by APCVD, Thin Solid Films. 221 (1992) 166–182.

[16] M.S.A. Ziabari, S.M. Rozati, N. Najafi, Transparent Conductive Oxide Thermoelectric (TCO-TE) Generators: Fabrication and Comparing of ZnO, SnO_2, Fluorine Doped Tin Oxide and Fluorine Doped Zinc Oxide Materials as TCO TE Generators, J. Nanoelectron. Optoelectron. 14 (2019) 1331–1338. https://doi.org/10.1166/jno.2019.2631.

[17] F.P. Fehlner, Thin films on glass for liquid crystal displays, J. Non. Cryst. Solids. 218 (1997) 360–367. https://doi.org/10.1016/S0022-3093(97)00243-3.

[18] D.W. Sheel, H.M. Yates, P. Evans, U. Dagkaldiran, A. Gordijn, F. Finger, Z. Remes, M. Vanecek, Atmospheric pressure chemical vapour deposition of F doped SnO2 for optimum performance solar cells, Thin Solid Films. 517 (2009) 3061–3065. https://doi.org/10.1016/j.tsf.2008.11.121.

[19] R.G. Gordon, J. Proscia, F.B. Ellis, A.E. Delahoy, Textured tin oxide films produced by atmospheric pressure chemical vapor deposition from tetramethyltin and their usefulness in producing light trapping in thin film amorphous silicon solar cells, Sol. Energy Mater. 18 (1989) 263–281. https://doi.org/10.1016/0165-1633(89)90042-7.

[20] D.S. Bhachu, M.R. Waugh, K. Zeissler, W.R. Branford, I.P. Parkin, Textured fluorine-doped tin dioxide films formed by chemical vapour deposition, Chem. - A Eur. J. 17 (2011) 11613–11621. https://doi.org/10.1002/chem.201100399.

[21] R.A. and G.H.F. E. Rädlein, Atomic force microscopy of coated glasses, Fresenius. J. Anal. Chem. 353 (1995) 413–418. https://doi.org/10.1007/s0021653530413.

[22] Y. Schiele, F. Book, C. Demberger, K. Jiang, G. Hahn, Co-Diffused APCVD Boron Rear Emitter with Selectively Etched-Back FSF for Industrial n-Type Si Solar Cells, in: Proc. 29th EU PVSEC, Amsterdam, The Netherlands, 2014: pp. 821–824. https://doi.org/10.4229/29thEUPVSEC2014-2AV.2.4.

[23] M. Kambe, M. Fukawa, N. Taneda, K. Sato, Improvement of a-Si solar cell properties by using SnO_2:F TCO films coated with an ultra-thin TiO_2 layer prepared by APCVD, Sol. Energy Mater. Sol. Cells. 90 (2006) 3014–3020. https://doi.org/10.1016/j.solmat.2006.06.008.

[24] D. Hocine, M.S. Belkaid, M. Pasquinelli, L. Escoubas, J.J. Simon, A. Moussi, Improved efficiency of multicrystalline silicon solar cells by TiO_2 antireflection coatings derived by APCVD process, 16 (2013) 113–117. https://doi.org/10.1016/j.mssp.2012.06.004.

[25] F. Yan, J. Jiang, X. Chen, S. Tian, K. Li, Synthesis and characterization of silica nanoparticles preparing by low-temperature vapor-phase hydrolysis of SiCl4, Ind. Eng. Chem. Res. 53 (2014) 11884–11890. https://doi.org/10.1021/ie501759w.

[26] F. P. Fehlner, LCD panel production, United States Patent Office, 2017.

[27] T. Nagata, T. Tanaka, K. Miyake, H. Kurotaki, S. Yokoyama, M. Koyanagi, Micron-size optical waveguide for optoelectronic integrated circuit, Jpn. J. Appl. Phys. 33 (1994) 822–826. https://doi.org/10.1143/JJAP.33.822.

[28] H. Nakano, M. Matsumoto, Y. Onose, K. Ohta, Metal strain gauge pressure sensor using SiO_2/Si3N_4 diaphragm, IEEJ Trans. Sensors Micromachines. 139 (2019) 63–68. https://doi.org/10.1541/ieejsmas.139.63.

[29] B.T.M.L. P. Chesworth, Method of coating glass, United States Patent, 1988.

[30] K.O. Davis, K. Jiang, D. Habermann, W. V. Schoenfeld, Tailoring the Optical Properties of APCVD Titanium Oxide Films for All-Oxide Multilayer Antireflection Coatings, IEEE J. Photovoltaics. 5 (2015) 1265–1270. https://doi.org/10.1109/JPHOTOV.2015.2437272.

[31] V.Y. Vassiliev, Void-free pre-metal dielectric gap-fill capability with CVD films for subquater- micron ULSI (Invited paper), in: 6th Intern. Dielectr. ULSI Multilevel Interconnect. Conf., Santa Clara, USA, 2015.

[32] M.M.R. Leitch, J. Li, D. Thach, F. Moghadam, Implications of Surface Contaminants in a BPSG Deposition and Reflow Process, MRS Online Proc. Libr. Arch. 315 (1993) 117–123. https://doi.org/https://doi.org/10.1557/PROC-315-117.

[33] J. Kim, S. Hwang, J. Yi, SiO_2 films deposited at low temperature by using APCVD with TEOS/O_3 for TFT applications, J. Korean Phys. Soc. 49 (2006) 1121–1125.

[34] Y. Ikeda, Y. Numasawa, M. Sakamoto, Ozone/organic-source APCVD for conformal doped oxide films, J. Electron. Mater. 19 (1990) 45–49. https://doi.org/10.1007/BF02655550.

[35] S. Sivoththaman, P. De Schepper, W. Laureys, J.F. Nijs, R.P. Mertens, Improving low-temperature APCVD SiO_2 passivation by rapid thermal annealing for Si devices, IEEE Electron Device Lett. 19 (1998) 505–507. https://doi.org/10.1109/55.735761.

[36] A. Aguilar, S.Y. Herasimenka, J. Karas, H. Jain, J. Lee, K. Munoz, L. Michaelson, T. Tyson, W.J. Dauksher, S. Bowden, Development of Cu plating for silicon heterojunction solar cells, in: 2017 IEEE 44th Photovolt. Spec. Conf. PVSC 2017, Portland, OR, USA, 2017: pp. 1–4. https://doi.org/10.1109/PVSC.2017.8366440.

[37] D. Adachi, T. Terashita, T. Uto, J.L. Hernández, K. Yamamoto, Effects of SiO_x barrier layer prepared by plasma-enhanced chemical vapor deposition on improvement of long-term reliability and production cost for Cu-plated amorphous Si/crystalline Si heterojunction solar cells, Sol. Energy Mater. Sol. Cells. 163 (2017) 204–209. https://doi.org/10.1016/j.solmat.2016.12.029.

[38] J. liang Zhang, J. Li, L. ma Luo, Y. hua Wo, Microstructure and morphology of SiO_x film deposited by APCVD, J. Alloys Compd. 469 (2009) 535–538. https://doi.org/10.1016/j.jallcom.2008.02.014.

[39] T. Fujimoto, K. Okuyama, S. Yamada, M. Adachi, Effect of cluster/particle deposition on atmospheric pressure chemical vapor deposition of SiO_2 from four gaseous organic Si-containing precursors and ozone, J. Appl. Phys. 85 (1999) 4196–4206. https://doi.org/10.1063/1.370331.

[40] J.D. Chapple-Sokol, C.J. Giunta, R.G. Gordon, A Kinetics Study of the Atmospheric Pressure CVD Reaction of Silane and Nitrous Oxide, J. Electrochem. Soc. 136 (2019) 2993–3003. https://doi.org/10.1149/1.2096390.

[41] K. Okuyama, D. Huang, J.H. Seinfeld, N. Tani, Y. Kousaka, Aerosol formation by rapid nucleation during the preparation of SiO_2 thin films from $SiCl_4$ and O_2 gases by CVD process, Chem. Eng. Sci. 46 (1991) 1545–1560. https://doi.org/10.1016/0009-2509(91)87004-V.

[42] M. Adachi, K. Okuyama, M. Shimada, N. Tohge, J.I. Satoh, M. Muroyama, Gas-phase nucleation in an atmospheric pressure chemical vapor deposition process for SiO_2 films using tetraethylorthosilicate (TEOS), Jpn. J. Appl. Phys. 31 (1992) L1439–L1442.

[43] J. VanOmmeren, Silane Gas Venting/Treatment Lessons Learned and Best Practices, Process Saf. Prog. 30 (2011) 104–114. https://doi.org/https://doi.org/10.1002/prs.10414.

3 Novel approach to SiO_{2-x} APCVD system

3.1 Introduction

This chapter focuses on establishing a new method of APCVD SiO_{2-x} films suitable for solar cell applications. Further tackling of this method on other film materials e.g. TiO_2 and applications are still an option. A brief review of well-known gas precursors and deposition conditions to prepare SiO_2 films under atmospheric pressure are given. New gas precursors to replace the widely used ones are introduced for the first time in APCVD SiO_{2-x} utilized solar cells. After summarizing their properties, the precursors' reaction-kinetics and their relevant deposition mechanism are explained, theoretically. To overcome the drawbacks of conventional APCVD SiO_2 films mentioned in chapter 2, a newly developed reactor configuration, characterized by low cost and thermal budget, is demonstrated. The stability of system operation and reproducibility of products are attempted. Detailed characterization of the material quality is studied in chapter 5.

3.2 Newly developed SiO_{2-x} APCVD system by the hydrolysis of $SiCl_4$

In this section new reactant precursors are researched so as to save costs and simplify the SiO_2/SiO_x deposition process, whereby fewer safety precautions and pre-preparation steps are required opposing the state-of-the-art APCVD methods [1–4]. The major disadvantages of conventional SiO_2 APCVD technologies to overcome with the new method were the complexity of the deposition systems, the use of flammable reactant precursors, and the high thermal budget required for reactants dissociation and oxidation. The more sensitive APCVD downside – the dust formation due to mass transfer reaction, also known as gas-phase or molecular-diffusion reaction, was addressed to decrease the periodic cleaning and maintenance of the apparatus and prevent particle precipitation upon the deposited films. Elimination of impurity concentrations trapped in the deposited films as well as advances in the dielectric and chemical characteristics are studied in chapter 5.

The chemical precursors researched in this study are essentially silicon tetrachloride and water vapors. Excluding for PV applications the reaction between H_2O and $SiCl_4$ vapors, under atmospheric pressure, is often employed for other state-of-the-art applications, e.g. to generate silica nanoparticles at a high temperature of 1800 °C [5]. $SiCl_4$ vapor hydrolysis is also possible at 150 °C to form oxychloride particles, which are then converted to nearly monodisperse silica spheres through a further hydrolysis at 1000 °C. Si-O/Si-OH nanoparticles are usually synthesized as illustrated in the following sequential reactions [6]:

$$SiCl_4 + H_2O \rightarrow SiCl_3OH + HCl \tag{3.1}$$

$$SiCl_3OH + H_2O \rightarrow SiCl_2(OH)_2 + HCl \tag{3.2}$$

$$SiCl_2(OH)_2 + H_2O \rightarrow SiCl(OH)_3 + HCl \tag{3.3}$$

$$SiCl(OH)_3 + H_2O \rightarrow Si(OH)_4 + HCl \tag{3.4}$$

$$Si(OH)_4 + Si(OH)_4 \rightarrow (OH)_3\,Si\text{–}O\text{–}Si(OH)_3 + H_2O \tag{3.5}$$

$$Si(OH)_3\text{–}O\text{–}Si(OH)_3 + Si(OH)_4 \rightarrow Si(OH)_3\text{–}O\text{–}Si(OH)_2\text{–}O\text{–}Si(OH)_3 + H_2O \tag{3.6}$$

APCVD SiO_4/SiO_x by means of $SiCl_4$ hydrolysis at room temperature is formed as a porous-rough underlayer mainly applied to glass substrates for the purpose of supporting later deposition of antireflection or hydrophobic coatings [7]. APCVD SiO_2 layers made of $SiCl_4/H_2O$ can be executed at room temperature to increase the mechanical stability and optical properties of colloidal crystal lattices made of mono-disperse micro-spheres of silica, wherein their pore-size are controlled, hence their connectivity [8]. APCVD SiO_x, prepared this way, are applied as anti-humidity dual coating, together with TiO_x films, to act as encapsulation of phosphorous particles used in electroluminescence devices [9]. These coatings are attempted for the prolongation of the light emission life-span of the phosphorus particles under aggressive circumferential conditions (40°C, 90% humidity).

Our new APCVD method was developed to primarily deposit SiO_{2-x}/ SiO_x for PV applications as well as other optical applications by means of the hydrolysis of $SiCl_4$ vapor under atmospheric conditions. Detailed information is to be published elsewhere. The equation governing this exothermic reaction is as follows:

$$SiCl_4 + 2\ H_2O \rightarrow SiO_2 + 4\ HCl \qquad (3.7)$$

3.2.1 SiO_{2-x} APCVD system configuration

The newly developed SiO_{2-x} APCVD setup configuration illustrated in Figure 3.1 is constituted as follows:

1) Two evaporators made of plastic or glass containers with a volume of 1-2 liters used to vaporize the reactant precursors – the volatile $SiCl_4$ and H_2O, respectively. The evaporators are tightly connected to an inert gas source (utilized as a carrier gas) such as dry N_2 and to APCVD injector inlets; either directly for the $SiCl_4$ line or through another container, utilized as a humidity mixer, for the H_2O line.
2) A tight humidity mixer, which is the most important component, made of plastic or glass container with a volume of 1 – 2 liters, wherein an admixed dry N_2 is utilized to dilute the humid stream driven from the H_2O evaporator. This arrangement should remain at a constant temperature, for which a thermal insulation could be used.
3) A thermostat equipped with de-ionized (DI) water tank and a dual cooling/heating system for controlling and retaining the H_2O evaporator's temperature during the APCVD system operation. Another thermostat is optionally operated to control the temperature of the deposition chuck, between them an adjusted-temperature DI water stream is circulated. More details about the design of deposition chuck and the control of its temperature are elucidated in the next chapter.
4) A sealed chamber made of transparent polycarbonate (PC) plastic for laboratory scale-levels, wherein the relevant APCVD injector and $SiCl_4$ evaporator are mounted. It is supplied with an adjustable exhaust opening as well as adapted openings used to attach multiple pipe lines and electronics. An external transfer room made of the transparent PC plastic is attached to the chamber to enable placing and replacing processed samples and any equipment needed during the experiments. It is supplied with two sealed doors opened alternatively and thus

protect from escaping the aggressive gases from the chamber. In order to assure safe operation, two rubber gloves are fixed tightly to the front side of the chamber, which is in turn placed in a fume-hood used when cleaning the equipment for better security.

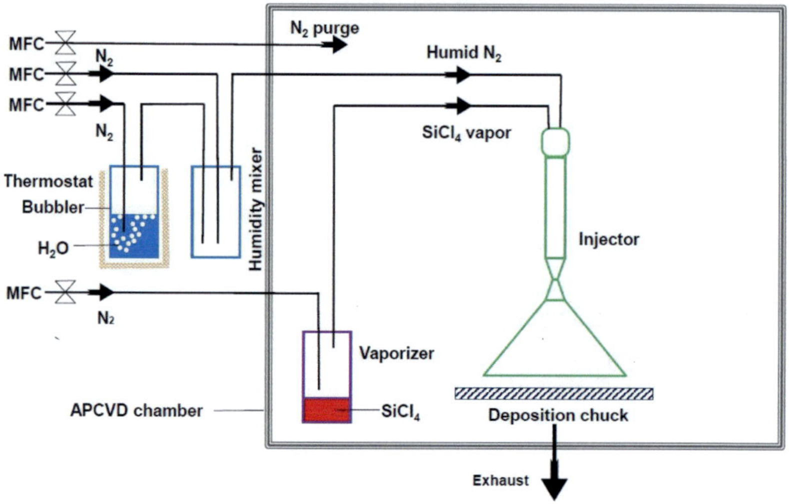

Figure 3.1: A diagram of the newly developed SiO_{2-x} APCVD system layout employed for PV applications

5) An APCVD injector made of thermoplastic materials by means of fused deposition modeling, or so-called three-dimensional (3D) printing technology. The employed 3D printing, utilizing polylactic acid (PLA) and acrylonitrile-butadiene-styrene (ABS) materials, enables customizing various injectors suitable for different utilities and variable substrate surface-areas. The injector's outlet is connected to the chamber's exhaust through an adapter featuring an adjustable opening. The detailed design of the APCVD apparatus and its relevant operating conditions are studied in the next chapter.
6) Pipelines insulated with Thermofoam or other variants so as to maintain a constant gas temperature throughout the system. They are employed to deliver the reactant gases between the sources and the injector inlets.

A lateral pipeline is installed to purge the APCVD chamber interior with dray N_2 gas prior to the deposition process.

7) Mass flow controllers (MFCs) used to precisely adjust and measure the flow rate of the gas stream in every pipeline – of the water bubbler, the humidity mixer, the $SiCl_4$ vaporizer and the purge stream.
8) Valves of 90° switch type installed at every gas stream and the main N_2 gas source (that is supplied with a pressure regulator) to open and close the pipelines when a quick control is needed as well as to calibrate the MFCs. These valves are beneficial for security purposes, as well e.g. if by unexpected occasion a shortcoming occurred, the relevant or main gas stream can be closed immediately.
9) Built-in humidity/temperature sensors of the kinds DHT11 and DHT22 (illustrated later in Figure 3.5) installed into the following positions: i) the water bubbler, ii) the humidity mixer, iii) the deposition chuck inside the injector, iv) the chamber volume, v) and the area surrounding the APCVD setup. Since the DHT11 sensor kind is unaffected by aggressive gases like HCl, known to be released in high amounts during the reaction, it is used inside the chamber and the injector. DHT22 sensors are sensitive to HCl so that, in order to avoid measurement interference, they are installed in the other setup components.
10) An Arduino microcontroller is employed to process the information coming from the built-in sensors and translate them to meaningful values –humidity and temperature of the interesting positions. Its terminals are connected on one side to the sensors through cables and on the other side to a computer, by which the relevant values are displayed. The microcontroller is programmed by means of the "C++" language. The code is written in such as way as to record the measured relative humidity (RH) and temperature (T) as well as the calculated absolute humidity (AH) and dew point (T_{dp}) of the media, with an interval of 5 seconds.
11) An endoscope camera, of the universal-serial-bus (USB) port kind, installed in the injector interior for counter-the-clock monitoring of i) film deposition progress, ii) accidental contamination, iii) and shortcoming such as deposition stop due to a precursor draining in the

evaporator. This is made possible by connecting the endoscope to a computer, with a suitable software a live display and even recorded video and captured pictures can be acquired for documentation.

12) A computer and its monitor utilized to operate the microcontroller and the endoscope camera. This allows for judging the deposition quality and controlling the operating conditions and parameters, by which customizing the SiO_{2-x} film properties and their deposition rates are facilitated.

It is worth noting that the tightly sealed APCVD chamber, in combination with by-products' exhaustion, are significantly the essential measures taken for safe usage of the newly developed SiO_{2-x} APCVD system. Therefore, its surrounding atmosphere is protected from any hazard or damage. Moreover, in order to prevent from corrosion (used to occur to metal components) with the by-product HCl gas, the whole system configuration is made of plastic materials, as well as with rubber and silicone sealing and caps. The APCVD setup can be upgraded to enable neutralizing the HCl gas when it is not available in the main exhaust of the laboratory.

3.2.2 $SiCl_4/H_2O$ deposition process

The operation of the newly developed SiO_{2-x} APCVD system includes, prior to the deposition process, simple pre-preparation steps as follows:

1) Purging of the APCVD chamber from humid air by dry N_2 gas, setting up maximum exhaust power, for about 20 min. This duration was found to reduce the RH inside the chamber volume to ~1 %, which is a prerequisite condition to prevent SiO_{2-x} formation in the chamber medium and deposition onto its inner walls. For a quick purging the dry N_2 gas is streamed at a pressure of 2-4 bars. After drying the chamber volume the exhaust is slowed down (by readjusting its relevant valve) and the pressure of N_2 purge is decreased to 1.5 bar, hence maintaining a slight under-pressure inside the chamber. For safety precautions, this prevents the aggressive gases ($SiCl_4$ and HCl) from getting into the laboratory's atmosphere. The under-pressure in the chamber can be readily indicated,

before and during the deposition, if the tightened gloves are sucked into its interior.

2) Adjusting the temperature of the deposition chuck, hence of the later mounted substrate. In most cases the deposition takes place at room temperature, thus the chuck temperature is maintained the same thereof. In special cases e.g. when SiO_{2-x} film characteristic is to be changed to porous structure (so called SiO_x), cooling of the deposition chuck is carried out. Heating of the deposition chuck is possible with the same cooling arrangement, as well.
3) Fine-tuning of the carrier gas flow-rate (Q) within the lines of humidity mixer, H_2O vaporizer, and $SiCl_4$ vaporizer. The flow rate, found to be adequate for cost-effective deposition, is in the range 0.5-2.5 l/min.
4) Fine-tuning the H_2O vaporizer's temperature, on the other hand, is performed after stabilizing its relevant flow rate by means of setting the thermostat temperature as required. The effective temperature needed during experiments was found to be in the range 20-29 °C. The temperature inside the $SiCl_4$ vaporizer and the humidity mixer, where the room-temperature N_2 stream collides with the humid N_2 stream driven from the H_2O vaporizer, are kept without any change (usually they are the same as the temperature of the APCVD chamber). This temperature profile is essential to avoid water condensation in the pipelines as well as upon the substrate surface, thus obstructing deficient oxide films characterized by bread-loafing and stiff structure.
5) Filling the two evaporators with DI-H_2O and the fluidized $SiCl_4$ 99.9%, after the set temperatures and flow rates are stabilized. The amount of $SiCl_4$ liquid dosed into the related evaporator is determined by many factors such as substrates' size and number, desired SiO_{2-x} film thickness, and flow rate of the N_2 carrier gas, hence resultant film deposition rate. After this step the substrate is loaded to the apparatus so that the deposition process starts.

The operation of the SiO_{2-x} APCVD system as well as the deposition process take place as follows: the DI-H_2O is evaporated due to the flowing N_2 carrier gas into the related evaporator, by which the gas stream is saturated with the water vapor. The humidified N_2 stream is then directed into the humidity mixer, wherein a precise AH value is calibrated thanks to the diluting N_2 flow introduced from the lateral stream line. It is worth noting that in this newly developed system AH is used as a measurement of the H_2O concentration in contrast with the so-far methods, where RH is the driving parameter [7–9]. The calibration of the humidity into the mixer is conducted by adjusting the flow rate of the two stream lines – humid N_2 driven from the water evaporator and admixed dry N_2. Subsequently, resultant humid mixture flows to its relevant APCVD injector inlet. Simultaneously, the $SiCl_4$ solution is vaporized and premixed with N_2 carrier gas into the related evaporator, wherein a precise $SiCl_4$ moiety concentration is tuned according to the N_2 flow rate passed through. Subsequently, the $SiCl_4$-premixed N_2 stream flows to its relevant APCVD injector inlet. The formulas and investigations to calculate all mentioned flow rates and other operating boundary conditions are explained later in this chapter. The two vapor streams collide within the APCVD injector and premix thoroughly until obtaining a uniform reactants' mixture prior to the substrate surface. The uniform mixture then continuously strikes and reacts with the substrate surface forming the SiO_{2-x} layer. After a deposition cycle is completed the chamber exhaust is opened to discharge the by-product HCl gas. Neutralizing the HCl gas can be executed by directing the exhaust line into a lateral water container prior to the disposal throughout the exhaust of the fume-hood. At first glance, one would assume inner wall contamination with SiO_{2-x} dust as a consequence of the well-known abrupt nature of reaction between $SiCl_4$ and H_2O. A novel solution to suppress the side effect of beforehand reactants' mixing is investigated in the next chapter.

The SiO_{2-x} films prepared with the newly developed APCVD system can be deposited upon plenty of substrate materials; by way of examples: all kinds of Si wafers, passivated metals, optic fibers, polymers, plastics, glass, etc. They can be applied onto numerous coating materials as well, e.g. SiN_x, AlO_x, a:Si, a:Si:H, TCO, etc.

3.3 Reactant properties and deposition boundary conditions

Silicon tetrachloride, also known as tetrachlorosilane, and water are inorganic compounds liquids ($SiCl_4$ is a volatile one), each of which, however, vaporizes under different conditions. The physical and chemical properties of these two species are prerequisite parameters for the prediction of their evaporation mechanism and hence their subsequent deposition process. This knowledge assists process information such as boundary conditions of the APCVD system operation (flow patterns and temperature profiles), reactant volume needed to replenish the vaporizers, and estimated film deposition rates. The reactant properties required to realize this information are e.g. vapor pressure or known as saturated vapor pressure (P_v), density (ρ), boiling point (T_b), heat of vaporization (ΔH_v), and molar mass (M). Table 3.1 depicts the value of these properties for both $SiCl_4$ and H_2O in the liquid phase [10,11].

Table 3.1: Physical and chemical properties of the liquid $SiCl_4$ and H_2O reactants. The P_v and ρ are measured at room temperature (25°C). The T_b and ΔH_v are measured at 1 atm [10,11].

Reactant	P_v (mmHg)	ρ (g.cm^{-3})	T_b (°C)	ΔH_v (kJ.mol^{-1})	M (g.mol^{-1})
$SiCl_4$	236	1.483	57.65	30.40	169.90
H_2O	23.8	0.997	99.98	40.65	18.015

3.3.1 Saturation vapor pressures

Since the evaporation of H_2O and $SiCl_4$ is a preeminent step of the SiO_{2-x} APCVD, its mechanism can be clarified by the relationship between their saturated vapor pressure and their temperature and intermolecular force.

Saturation vapor pressure is, by definition, the pressure of a vapor, which is in thermodynamic equilibrium with its liquid (or solid) when measured in a sealed container, wherein the substance exists in both liquid (or solid) and vapor phases [12]. Note that when the substance of interest is solid, this phenomenon is called sublimation instead of evaporation. However, the liquid vaporization takes place since some of its surface molecules are characterized by high kinetic energy enough to escape to the liquid headspace. Owing to the bouncing of escaped molecules in vapor phase, they collide with the liquid and reengage into its surface through a process known as condensation. Simultaneously, other high-energy molecules evaporate from the surface keeping the headspace occupied with actively bouncing molecules, known as saturation with vapor phase substance, hence a certain pressure exerted above the liquid surface. When the number of evaporated molecules is equal to that of the condensed ones, the equilibrium pressure is reached. This equilibrium pressure expresses the vapor pressure of the investigated liquid.

Two factors governing the saturation vapor pressure of a liquid are the intermolecular force between its surface molecules and the temperature. The stronger the intermolecular force, the lower the vapor pressure, and vice versa, the weaker this force, the higher the vapor pressure [13]. On the other hand, the higher the liquid temperature, the higher its vapor pressure. This relationship is expressed by various equations – the most common ones are:

Clapeyron-Clausius equation [12]:

$$log\, P_v = \frac{\Delta H_v}{(2.3)R\,.T} + constant \qquad (3.8)$$

where:

P_v is the vapor pressure (*Pa*).

ΔH_v is the heat of vaporization ($J.mol^{-1}$).

R is the universal gas constant ($R = 8.314\ J.mol^{-1}K^{-1}$).

T is the temperature (*K*).

and Antoine equation [14]:

$$ln\, P_v = A - \frac{B}{T + C} \qquad (3.9)$$

where:

P_v is the vapor pressure (*mmHg*).

T is the temperature (*K*).

A, B, C are the Antoine coefficients, defined by the investigated studied liquid. In this study they are summarized in Table 3.2 [10,11].

Table 3.2: Antoine coefficients for SiCl4 and H2O liquids [10,11]

Reactant	A (mm Hg)	B (mm Hg / K)	C (K)
$SiCl_4$	4.09777	1200	37
H_2O	8.140191	1810.94	244.485

Based on the above it is recommended to maintain the reactant evaporators employed in the new APCVD system at constant temperatures, e.g. using thermostats and insulations. This is to prevent variation in the subsequent evaporation rate on account of temperature change.

3.3.2 Evaporation rate

Evaporation rate of a liquid is the rate of its transformation to the vapor phase. It is also known as a mass or a volume that vaporizes from a liquid surface per unit time.

In practice the evaporation rate of a pure liquid subjected to an external flow of e.g. a carrier gas (the case of our evaporators) is influenced by many factors. Namely, these factors are: physical and chemical properties (P_v, ρ, T_b and ΔH_v), volatility, temperature, peripheral pressure, and surface area (exposed to the carrier gas) of the vaporizing reactant as well as velocity, temperature, and liquid-molecule concentration into the carrier-gas flow. Accordingly, prediction of the evaporation rate is possible with multiple expressions based on fundamental heat and mass transfer equations [14]. The expressions are disputed as to numerous aspects such as the carrier-gas material, and its humidity and flow characteristic (laminar or turbulent), hence planarity of the liquid surface. It is also necessary to know if the carrier gas is a pure material or a mixture of many gases. One of the expressions of the specific evaporation rate has been presented accordingly by Van den Bosch and Weterings, 2005 [15], as follows:

$$\acute{E} = K_m \frac{P_v \, . \, T_i \, . \, MW}{R \, . \, T_i} \tag{3.10}$$

where:

$\acute{E}$ is the specific evaporation rate (*g/m²s*).

K_m is the mass transfer coefficient (*m/s*).

P_v is the vapor pressure of the liquid chemical at its temperature (*atm*).

T_i is the temperature of the interface between the liquid chemical and the above gas flow (*K*).

MW is the molecular weight (or so called the molar mass) of the chemical (*g/kmol*).

R is the ideal gas constant ($R = 8.205\ m^3 \cdot atm/kmol.K$).

Another simple expression to predict the specific evaporation rate, considering the carrier-gas velocity characterized by laminar flow above a flat liquid surface, was used by EPA, 1999, based on investigation by et. al Mackay [16], as follows:

$$\acute{E} = \frac{2.2 \,.\, 10^{-4} \,.\, v_{cg}^{0.78} \,.\, MW^{0.667} \,.\, P_v}{T_i} \tag{3.11}$$

where:

v_{cg} is carrier-gas speed above the liquid surface (*cm/s*).

MW is liquid molecular weight (*unitless in this equation*).

When considering the surface area of the vaporized chemical precursor, the resulting evaporation rate is measured in (g/s) and calculated as follows:

$$E = \frac{2.2 \,.\, 10^{-4} \,.\, V_{cg}^{0.78} \,.\, MW^{0.667} \,.\, A_p \,.\, P_v}{T_i} \tag{3.12}$$

where:

E is the evaporation rate measured in (*g/s*).

A_p is liquid surface area (cm^2).

Based on the equations and theory mentioned above, it is intuited that the evaporation rate of a liquid increases with the increase of its temperature. This is due to the fact that the elevated temperature of a substance induces an increase of its relevant vapor pressure and hence vaporized volume per unit time. This is therefore consistent with the influence of the liquid substance, as though the greater the saturated pressure characterizing the liquid, the greater its volatility and hence the higher its rate of evaporation. The equation (3.11) reveals that the evaporation rate increases with the increase of gas velocity flowing upon the liquid surface. This holds true when the flow rate of the carrier gas is considered instead of the velocity. Due to its continuous flow, the carrier-gas stream is impeded from being saturated with the evaporated molecules, thus the equilibrium between the carrier gas and the liquid surface is prevented. Consequently, more volume of the liquid is permitted to evaporate. When the carrier-gas velocity (or flow rate) increases, its content of evaporated molecules per unit volume decreases, yet the ultimate rate of evaporation increases and vice versa. These causes and effects conform with the necessity of introducing a completely dry carrier gas so as to precisely control the evaporation rate of the intended liquid. The equation (3.12) depicts that the evaporation rate increases with the surface area of the liquid exposed to the external gas stream. This is due to the fact that the liquid spread on a wider area provides larger number of surface molecules characterized by high kinetic energy, hence higher evaporated volume of the liquid.

3.3.3 Evaporation study of $SiCl_4$

Since $SiCl_4$ evaporation is quite sensitive (due to its high vapor pressure) compared to the H_2O, our focus was on investigating the evaporation rates and reactant concentration resulted from a self-made $SiCl_4$ evaporator. $SiCl_4$ evaporation rates and mass concentration into the carrier gas were studied. Process stability over time was investigated so as to obtain a reproducible operating condition, hence subsequently reproducible film deposition process. The self-made $SiCl_4$ evaporator of wick kind [17] was constituted in such as way as to ensure stable evaporation rates, where the evaporated volume is constant all along the APCVD-system operation. For that, the aforementioned boundary conditions (P_v, T_i, V_{cg}, and A) were stabilized. A diagram of this evaporator is illustrated in Figure 3.2. The $SiCl_4$ evaporator comprises a glass or a plastic container, into which four glass beakers with a size of 10 ml are installed. The beakers featuring a diameter of

20 mm are loaded with cotton wool. The dense cotton wool with an arranged weight of 200 mg forms the so-called wick structure, owing to which a capillary action, also known as wicking, is promoted to the dosed $SiCl_4$ solution and therefore a continuous evaporation is attained. Additionally, since the wicking allows a constant liquid volume to reside upon the beaker-opening level, the saturation vapor pressure remains constant and hence a stable evaporation rate (when the other mentioned conditions are considered).

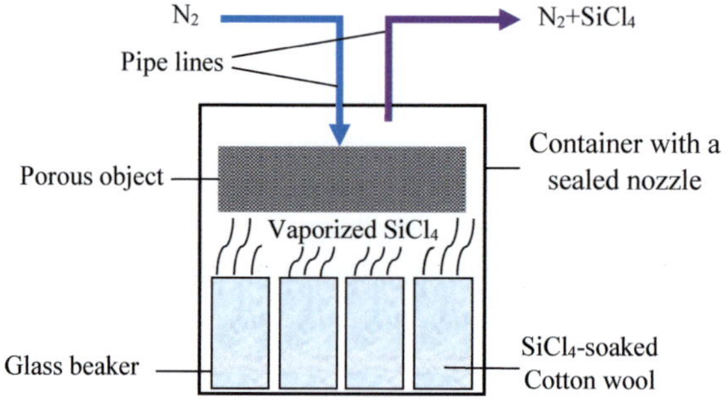

Figure 3.2: Diagram illustrates the $SiCl_4$ evaporator which includes 4 beaker containers loaded with cotton wool, porous object at the top to mix the resultant vapor uniformly, and two openings to insert and exert the carrier and working gases, respectively.

In order to transfer the evaporated volume to the APCVD injector, a sealed nozzle furnished with two openings (inlet and outlet) was fixed to the container. The inlet opening is attached to a porous object (e.g. porous stone or 3D-printed polymer sponge) spaced above the beaker openings, as though a uniform N_2 flow is distributed all over the container upper-portion. The evaporated $SiCl_4$ is therefore premixed with the N_2 carrier-gas homogeneously. This arrangement is an essential prerequisite for the uniform evaporation among the 4 beakers, simultaneously. The reason behind using multiple beakers featuring a relatively large length and small opening diameter is to precisely control the evaporation rate, as the smaller the exposed liquid surface, the lower the resultant evaporation rate. Consequently, it is possible to exploit the beakers individually when lower evaporation rate is necessitated. The outlet opening is, in turn, connected to the working gas line so that to transfer the N_2-$SiCl_4$ mixture to the relevant injector inlet.

A simplified experimental way was used in this study to measure the evaporation rate of a previously defined-mass/defined-volume liquid. It is to subtract the liquid mass or volume after evaporation from the original one and divide the difference, which refers to the vaporized amount, by the measured time. The calculation equation is as follows:

$$E_{SiCl4} = \sum_{n=1}^{\infty} \frac{\Delta v_{n(SiCl4)}}{\Delta t_E} = \frac{\Delta v_{n(SiCl4)} \cdot \rho_{SiCl4}}{\Delta t_E} \qquad (3.13)$$

Another expression of this equation is:

$$E_{SiCl4} = \frac{v_1 - v_2}{\Delta t_E} = \frac{m_1 - m_2}{\Delta t_E} \qquad (3.14)$$

where:

E_{SiCl4} is the evaporation rate from the $SiCl_4$ vaporizer (*mg/s*) or (*ml/s*). It refers to the mass flow of $SiCl_4$ reactant through the relevant line, as well.

$\Delta v_{n(SiCl4)}$ is the difference in the reactant volume within the vaporizer (*ml*).

ρ_{SiCl4} is the respective density of the evaporated $SiCl_4$ reactant (*g/ml*).

v_1 and v_2 are the reactant volumes measured before and after evaporation, respectively (*ml*).

m_1 and m_2 are the reactant masses measured before and after evaporation, respectively (*mg*).

Δt_E is the measured evaporation period (*s*).

In order to determine the appropriate flow rate required for delivering the exact concentration of $SiCl_4$ molecules, at which a stoichiometric reaction is realized with the H_2O molecules, it is necessary to making use of the measured $SiCl_4$ evaporation rate. The following equation, which is concluded according to equation (3.12), shows the correlation between these operation parameters:

$$Q_{SiCl4} = \frac{E_{SiCl4}}{Kc_{\ (SiCl4)}} \qquad (3.15)$$

where:

Q_{SiCl4} is the prerequisite flow rate of the carrier gas into the $SiCl_4$ evaporator (*l/s*).

E_{SiCl4} is the resultant evaporation rate from the $SiCl_4$ evaporator ($SiCl_4$ mass flow) (*mg/s*).

$Kc_{(SiCl4)}$ is a calculated mass concentration of $SiCl_4$ molecules into the carrier-gas stream based on the desired $SiCl_4$-to-H_2O molar ratio (mg/m^3).

Experimental investigations were performed to fine-tune the $SiCl_4$ evaporation rate into the self-made evaporator, by which no excessive chemicals are introduced to the APCVD injector, hence cost-effective operation of the deposition system and ultimately reliable film quality. The evaporation characteristics were tackled to be as follows: i) the carrier gas is as dry and pure as N4.0 which equals 99.99 % purity, ii) the liquid vaporized is an individual material of pure $SiCl_4$ (99.9 % purity), iii) the $SiCl_4$ liquid top-surface is, due to the wicking, of a constant level and the interface with the upper flow is of a constant temperature of 25 °C, hence a constant vapor pressure, iv) and the flow of the N_2 carrier gas is laminar where its rate is in the range 0.02 – 2.4 l/min.

The four wick beakers of the evaporator were dosed, each with 2 ml $SiCl_4$, after scaling the evaporator empty. The evaporator was then mounted into the APCVD chamber set under a constant temperature of 25 °C. This was made possible by means of directing a halogen-light beam with a predetermined distance toward the APCVD chamber. Three flow rates of the N_2 carrier gas, to be examined, were setup as such to assure a laminar flow into the evaporator. To measure the evaporation rate, the reduction in the $SiCl_4$ mass released from the evaporator was measured every 2 minutes using a highly sensitive scale. The standard deviation of the measured evaporation rates was calculated so as to assess the stability of the evaporation versus time. The equation to calculate the standard deviation is as follows:

$$\sigma = \sqrt{\frac{1}{N}\sum_{n=1}^{N} x_n^2 - \bar{x}^2} \qquad (3.16)$$

where:

x_n is the measured evaporation rate every 2 minutes.

$\bar{x}$ is the average rate.

N is the number of measured rates during the whole period of time.

The experiments were repeated numerous times with various $SiCl_4$ volumes to rest assured the reproducibility. The time periods of draining $SiCl_4$ volumes from the evaporator is possible to predict by means of using the equation (3.13) or (3.14).

The measured mass reduction of the $SiCl_4$ amount was consistent, for longer than 35 min, at 0. 730, 1.24 and 1.36 g/s at the flow rates 0.14, 0.35 and 0.78 l/min, respectively. The interpolation fit of the plots was linear as shown in Figure 3.3 which indicates the stable evaporation rate versus time. The calculation of standard deviation resulted in 0.049 evidencing a quite low change in the evaporation rates during the chosen period of time. It is, as well, seen that the slope of the linear plots increased gradually with the increased flow rates to be 0.78, 1.24, and 1.37, at 0.14, 0.35 and 0.78 l/min, respectively. Practically, this proves the proportional relationship between the evaporation rate and the set flow rate of the carrier gas.

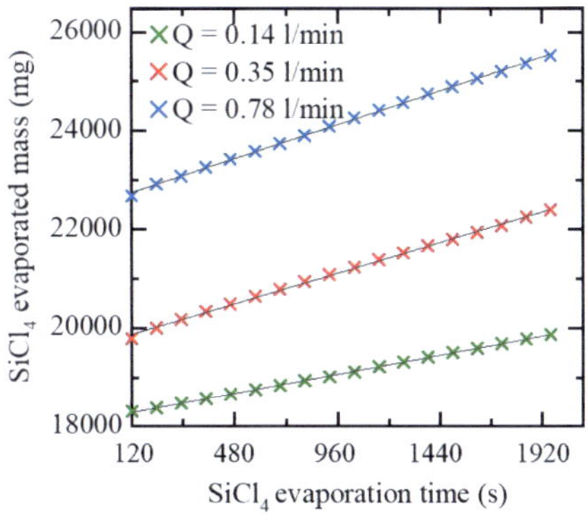

Figure 3.3: A graph showing the evaporated mass of $SiCl_4$ as a function of time at three gradual flow rates and a constant temperature of 25 °C. The linear fit of the plots shows a partially stable evaporation rate for longer than 2000 s. The time axis starts at 120 s as from which the linear plot fitting can be applied.

When it comes to $SiCl_4$ reactant, the governing parameter to a precise control of the SiO_{2-x} deposition rate and reaction stoichiometry was found to be its concentration in the N_2 stream. For that, investigations of the resultant $SiCl_4$ concentration into the N_2 stream were performed at gradual N_2 flow rats. Three sets of experiments were repeated under the same boundary conditions so as to prove the reproducibility as well as the self-made evaporator reliability. The constant parameters were evaporator temperature of 25°C and one-time dosed $SiCl_4$ volume of 2.5 ml in each wick beaker. For each set of experiments, laminar flows of the N_2 carrier gas were adjusted ranging from 0.21 to 2.0 l/min. the evaporation rates were measured using the previous manner. Consequently, the mass concentration of the $SiCl_4$ flowed into the N_2 stream was calculated using the equation (3.15) and the results were plotted in comparison with the corresponding evaporation rate. As depicted in Figure 3.4, apart from the additional evidence on the proportional relationship between Q and E_{SiCl4}, the results were reproducible, among the three experiment sets, to a further extent with a low margin of error. Consequently, the determination of $SiCl_4$ mass concentration driven to the injector is precisely realizable. It is worth noting that the increase of E_{SiCl4} is not linear with the Q which complies with the equations (3.11) and (3.12).

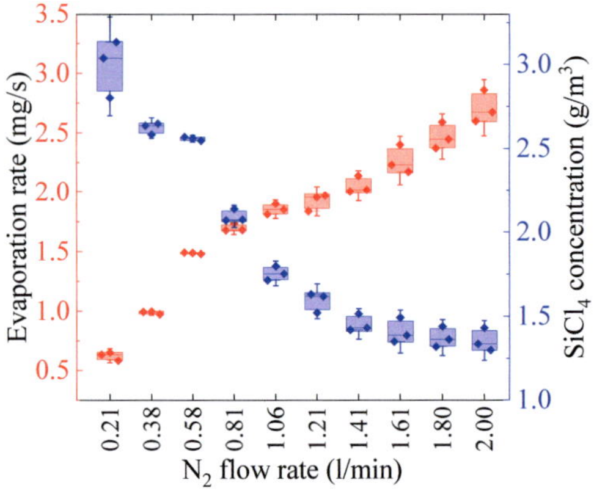

Figure 3.4: A graph showing the $SiCl_4$ evaporation rate and its concentration within the N_2 stream as a function of N_2 flow rate passed through the respective evaporator. The values represent three-time measurements at a constant evaporator temperature of 25° C.

3.3.4 Evaporation study of H_2O

In this section the evaporation mechanism of the DI-water produced to interact with $SiCl_4$ is investigated. An outstanding approach was established to accurately control the H_2O vapor concentration into the N_2 stream as well as its evaporation rate. Calculation of the parameters influencing H_2O evaporation were briefly introduced. An experimental investigation of the H_2O evaporation rate and mass concentration were performed according to the evaporator temperature and the carrier-gas flow rate.

The DI-water vaporizer is a bubbler arrangement comprising a bubbling object of e.g. porous stone completely submersed in the filled DI-water volume. The bubbling object is attached to an inlet opening imbedded into the bubbler nozzle, which is in turn attached to a dry N_2 flow-line. When the dry N_2 flows through the bubbling object it is distributed all across the water volume, thus the H_2O vapor is produced copiously. Multiple bubblers are optionally used for improved evaporation. The vapor-saturated media usually features an RH of 100% and is transferred through an outlet opening into the vaporizer nozzle. A vapor-saturated line is designated to deliver the H_2O vapor to the humidity mixer. The humidity mixer is a container without an incorporated porous object or liquid H_2O. However, it is an essential unit to adjust the AH, is to be introduced to the APCVD injector, as it is infeasible in the vaporizer itself. The humidity mixer is supplied with multiple inlets as well as a single outlet imbedded into the respective nozzle. One inlet is designated for the dry N_2 stream responsible for adjusting an accurate AH. The other inlets are designated for the moist media driven from the vaporizer to be diluted in the dry N_2. The N_2-H_2O premix is performed sufficiently owing to the injection position of the fresh gases at the bottom of the mixer unit. A so-called working vapor line is attached to the mixer outlet-opening so as to deliver the preadjusted AH stream to the relevant inlet of the APCVD injector. Figure 3.5 illustrates a diagram of this evaporation arrangement. In order to sufficiently obtain an accurate control of the H_2O vaporization it is beneficial to investigate the humidity conversion influences and formulae. For that, a clarification is required about the relative humidity, the dewpoint, and the absolute humidity [18]. Relative humidity is known to be a way to measure the presence of water vapor based on pressure. By definition it is the ratio of the partial water vapor pressure to its saturation vapor pressure at the holding gas temperature and pressure. It expresses the actual amount of water vapor, present in a defined gas volume, in comparison with the maximum amount that is possible to be present in this volume at the same temperature. Therefore, it reaches 100% when the actual amount of water vapor is equal to the maximum amount possible, wherein the holding gas is saturated with water vapor.

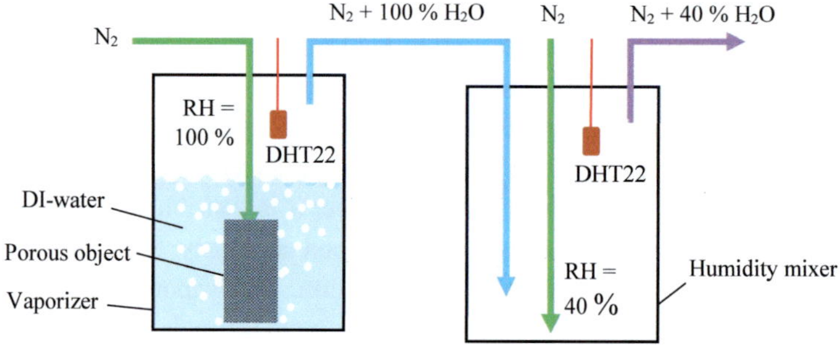

Figure 3.5: Diagram demonstrates the H_2O evaporation arrangement incorporating a DI-water vaporizer (**left**) and a humidity mixer (**right**). Built-in humidity/temperature sensors of the DHT22 kind are installed at the top of the two containers.

The relative humidity is expressed using the following equation:

$$RH = \frac{P_w}{P_{ws}} 100 \tag{3.17}$$

where:

RH is the relative humidity (*%*).

P_w is the partial water vapor pressure (*Pa*).

$P_{ws} = P_v$ is the saturation vapor pressure of water (*Pa*).

As mentioned previously, the saturation vapor pressure P_v is temperature dependent. This, therefore, holds true for the relative humidity as well. The higher the holding medium temperature, the higher the saturation water vapor pressure, and hence the lower the relative humidity considering that the partial pressure of the water vapor is unchangeable (the vapor amount is constant).

Another way to measure the presence of water vapor in a medium is by the dewpoint temperature. It is the temperature at which condensation of the water vapor, present in 100% RH medium, begins to occur. This takes place either by cooling down the medium featuring RH of 100% or by introducing more water vapor to the saturated medium (e.g. in the water bubbler), wherein the actual medium temperature is identical to the dewpoint one. The dewpoint temperature is calculated in correlation with RH from the following equation:

$$T_{dp} = \frac{T_n}{\left[\frac{m}{log_{10}(\frac{P_w}{A})} - 1\right]} \quad (3.18)$$

where:

T_{dp} is the dewpoint temperature (°*C*).

T_n, m, and A are constants of water taken from Table 3.3.

Table 3.3: Constants for the calculation of P_w and T_{dp} of water. From [18].

A (Pa)	m (Unitless)	T_n (°C)	Max error (%)	Temperature-range (°C)
6.116441	7.591386	240.7263	0.083	-20...+50
6.004918	7.337936	229.3975	0.017	+50...+100
5.856548	7.27731	225.1033	0.003	+100...+150
6.002859	7.290361	227.1704	0.007	+150...+200
9.980622	7.388931	263.1239	0.395	+200...+350
6.089613	7.33502	230.3921	0.368	0...+200

The operation of the newly developed SiO_{2-x} APCVD system is sensitive to condensation throughout the H_2O vapor line as well as upon the substrate surface. Condensation of the H_2O vapor and precipitation onto the inner walls of the transferring pipes reduce the molecules' number essential for a stoichiometric reaction, thus the $SiCl_4$ are excessive and unexploitable. Even worse, if the H_2O molecules condensate upon the substrate surface, it was found that the deposited SiO_{2-x} film is extremely cracked, hence unreliable film functionality (more details are demonstrated in chapter 5). This gives rise to the precise control of the temperature of both H_2O flow lines and substrate surface, by which T_{dp} is avoided all along.

In practice, it is realizable to increase the actual amount of water vapor produced in the bubbler by means of increasing the specific volume of the holding gas. This is, in turn, made possible by means of increasing the bubbler temperature at a constant RH of 100%. The mass of water vapor held in a defined volume of a dry gas is known as the absolute humidity (also known as the water concentration), which refers to the actual amount of water vapor. The absolute humidity into the gas volume is calculated using one of the following equations:

$$AH = \frac{m_{H2O}}{v_{mix}} \tag{3.19}$$

$$AH = D\,\frac{P_w}{T_b} \tag{3.20}$$

where:

AH is the absolute humidity or so called the water vapor concentration into a unit volume of dry gas (*g/m³*).

m_{H2O} is the weight of the water vapor held in the volume of dry gas (*g*).

v_{mix} is the volume of the mixture of water vapor and its holding gas (*m³*).

D is a constant = 2.16679 (*gK/J*).

T_b is the temperature of the bubbler including the water vapor (*K*).

P_w can be derived from the equation (3.18), considering that the T_b equals T_{dp} as the relative humidity in the bubbler (RH_b) is 100%, as follows:

$$\frac{m}{\log_{10}\left(\frac{P_w}{A}\right)} - 1 = \frac{T_n}{T_b}$$

$$log_{10}\left(\frac{P_w}{A}\right) = \frac{m}{1+\frac{T_n}{T_b}}$$

$$\frac{P_w}{A} = 10^{\wedge(\frac{m.Tb}{Tn+Tb})}.100 \quad (3.21)$$

Theses humidity parameters can be extracted from the psychrometric chart shown in Figure 3.6, as well.

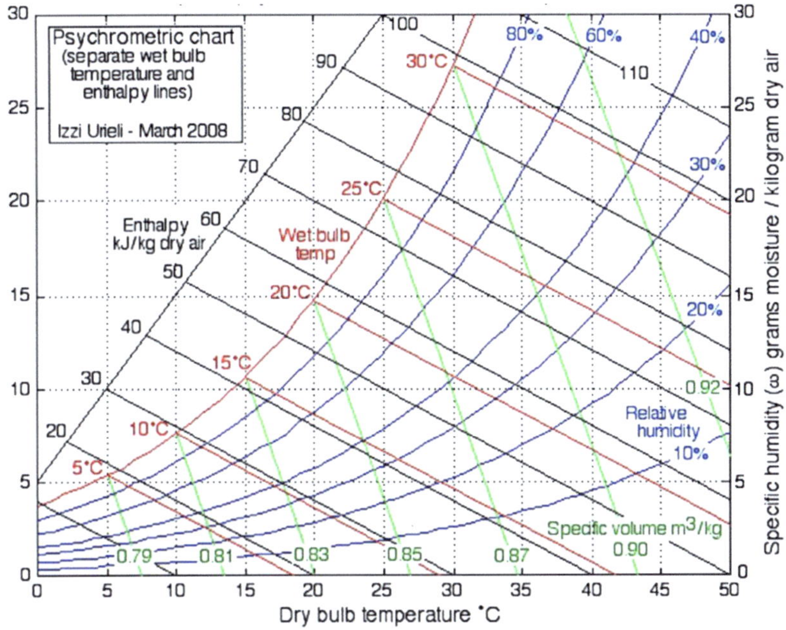

Figure 3.6: A simplified psychrometric chart at the ambient pressure. Note that the enthalpy lines are overlapped with the wet bulb temperature ones. From [19].

Based on the above, when it comes to our H_2O evaporation arrangement, RH at the outlet of the humidity mixer can be derived as follows:

$$RH_m = \frac{AH_m}{AH_b} 100 \quad (3.22)$$

where:

RH_m is the alternative humidity at the outlet of the humidity mixer (*%*).

AH_m and AH_b are the H_2O vapor concentration into the humidity mixer and the water bubbler, respectively (*g/m³*).

RH_m can be calculated based on the flow rates into the water bubbler and the humidity mixer, as well. The expressive equation is:

$$RH_m = \frac{Q_b}{Q_b + Q_m} 100 \tag{3.23}$$

where:

Q_b is the flow rate of the N_2+H_2O mixture flowing through the water bubbler (*l/min*).

Q_m is the flow rate of the dry N_2 flowing through the humidity mixer (*l/min*).

A comparative investigation between the RH of humidity mixer measured by the DHT22 sensors and the one calculated using equation (3.23) was performed for two different flow rates (0.35 and 1.06 l/min) of dry N_2 into the mixer. The experiments were carried out at gradual flow rates from 0.02 to 2.35 l/min into the water bubbler. The measured temperatures in the bubbler and the mixer were constant at 22 and 24.5 °C, respectively. An advantage of this investigation was to verify the accuracy of the utilized DHT22 sensors and hence the correspondence with the calculated values. The relationship between the flow rate into the H_2O bubbler and the relative humidity into the humidity mixer was investigated, as well. The results were consequently plotted as shown in Figure 3.7.

The interpolated plots in Figure 3.7 (left) show a reasonable match between calculated and measured RH_m irrespective of the Q_m or Q_b set values. It was therefore deduced that the DHT22-measured RH_m values are reliable enough to subsequently conclude the H_2O concentration and evaporation rates into the working gas stream (N_2+H_2O mixture). RH_m was shown in Figure 3.7 (right) to increase non-linearly with the increase of Q_b at constant bubbler and mixer temperatures. Because of the increase in Q_b the P_w into the bubbler increases towards the value P_{ws} (the vapor maximum pressure at the studied temperature), by which RH_b is elevated as clarified in equation (3.17), hence higher RH_m. On the other hand, it was shown that the lower the Q_m, the higher the resultant RH_m. This is ascribed to the less admixed dry N_2 volume in the mixer, due to which less dilution of the humidity is induced, thus high H_2O content into the N_2 gas is obtained.

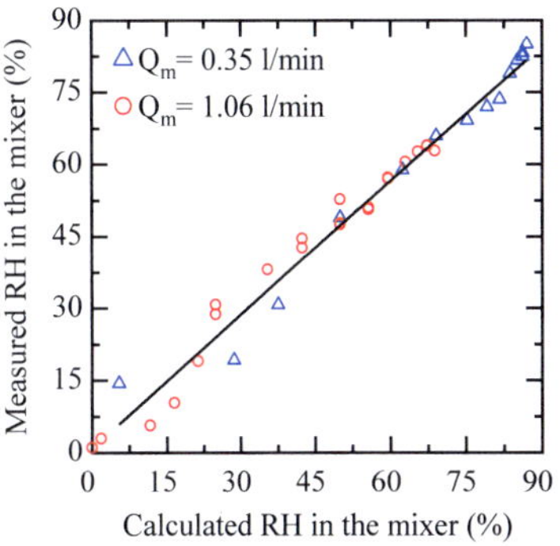

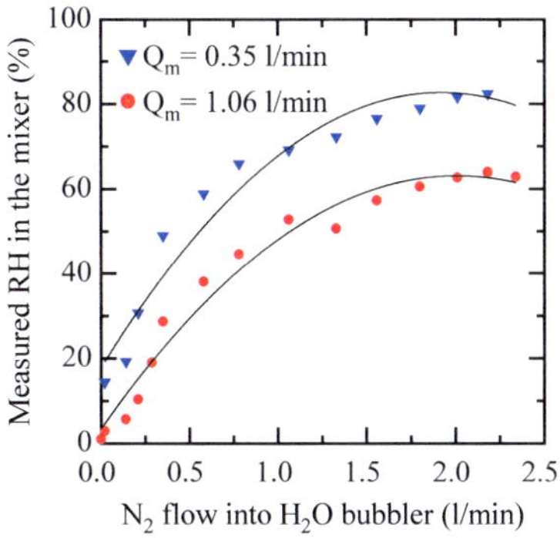

Figure 3.7: Left: A graph showing nearly linear relation between calculated and measured RH in the humidity mixer at a gradual flow rates into the bubbler for two different flow rates of dry N_2 into the humidity mixer. **Right:** A graph showing a polynomial change of the RH in the humidity mixer as a function of the flow rate into the H_2O bubbler at two different N_2 flow rates in the humidity mixer.

Since it is temperature dependent, RH is insufficient for gauging the actual H_2O amount necessitated for designing a stoichiometric reactants' mixture. Consequently, it is, instead, beneficial to consider the AH when studying the evaporation with the aforementioned vaporizer arrangement. AH is, in this case, calculated with the assistance of equations (3.22) and (3.23). Unlike the $SiCl_4$ evaporation, the term "evaporation rate" will be replaced by the term "mass flow" when it comes to H_2O evaporation study. In order to determine the total flow rate of the N_2+H_2O mixture at the humidity mixer outlet the H_2O vapor concentration and its mass flow are considered similar to equation (3.15) as follows:

$$Q_{(b+m)} = \frac{U_m}{AH_m} = \frac{U_b}{AH_m} \tag{3.24}$$

where:

$Q_{(b+m)}$ is the total flow rate at the humidity mixer outlet (*l/min*).

U_b and U_m are the mass flow of the H_2O vapor through the bubbler and the mixer, respectively (*mg/s*). They will be proved to be equal in a later inspection.

Another expression correlating with the flow rates is as follows:

$$Q_{(b+m)} = Q_b + Q_m \tag{3.25}$$

The flow rate into the bubbler can be estimated from the equation:

$$Q_b = \frac{U_b}{AH_b} = \frac{U_m}{AH_b} \tag{3.26}$$

The flow rate of dry N_2 through the humidity mixer is therefore calculated using equation (3.24).

An experimental investigation of the Q_b and Q_m impacts on the AH_{m}, introduced to the APCVD injector, was executed so that a precise adjustment of H_2O mass concentrations is to be obtained when operating the SiO_{2-x} APCVD system. Cardinally, RH and T were measured in the water bubbler and the humidity mixer with the DHT22 sensors, whereby three Q_m values of 0.35, 1.06 and 1.80 l/min were setup. For each Q_m the RH and T were measured at elevated Q_b values from 0 to 2.5 l/min. The temperatures were retained at 22.5 and 24 °C in the bubbler and the mixer, respectively. With the assistance of the computing system, incorporating Arduino microcontroller and provided C-programmed code, the AH measurements were processed and displayed using an accompanying computer. The results were plotted as shown in Figure 3.8. The AH_m was found to increase polynomially with the Q_b at the same Q_m and temperature. This is due to the increase of RH_b as illustrated in Figure 3.7 (right). The reason behind this curve behavior is that at the same temperature AH value is unincreasable after the one corresponds with RH = 100 %, regardless of higher N_2 flow rates. This occurs because no more H_2O can be incorporated in a unit volume of dry N_2. On the other hand, due to dilution in the dry N_2 within the humidity mixer, this H_2O concentration diminishes along with the increase of the respective N_2 flow rate. This effect of the humidity mixer was found to be significant when fine tuning of AH_m values (of e.g. 0.1 g/m^3) are to be set since it was found impossible to precisely adjust the H_2O concentration right into the water bubbler.

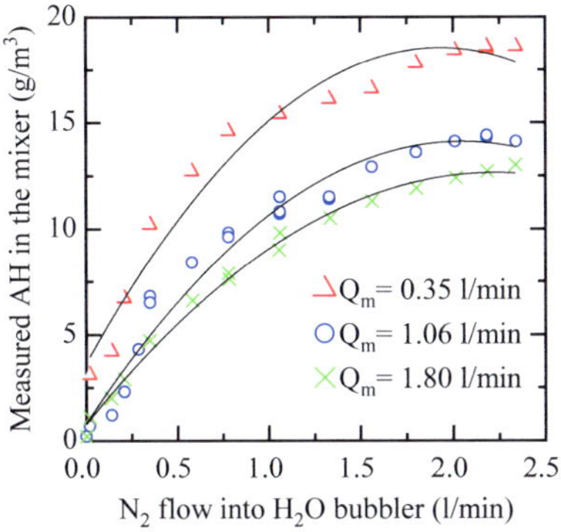

Figure 3.8: A graph depicting the measured H_2O concentration at the mixer outlet as a function of N_2 flow into the bubbler at three different flow rates of dry N_2 and constant temperatures of the bubbler and mixer of 22.5 and 24 °C, respectively

Numerous experiments were carried out to investigate the impact of temperature and flow rates upon the H_2O mass flow measured at the humidity-mixer outlet, whereby the flow of N_2+H_2O mixture is controlled prior to the SiO_{2-x}APCVD injector. The MFCs were first calibrated so as toassure accurate measurements. The humidity mixer was maintained at a constant temperature in order to simulate the usual operation conditions of the SiO_{2-x} APCVD system. T, RH, AH and U were measured/calculated in the bubbler and the humidity mixer every 5 seconds. In one experiment set a variation of Q_b at various T_b ranges was conducted to inspect their influence on the U_b and U_m. The Q_b was set at0.58, 0.81, 1.8 and 2.43 l/min, sequentially. At every Q_b the T_b was changed with a random gradient (between 20 and 28 °C). Q_m was set to zero, thus the flow rate at the mixer outlet $Q_{(b+m)}$ was equal to Q_b. U_b and U_m were subsequently calculated considering AH_b and AH_m measured values. In another experiment set the change of U was measured at the mixer outlet at gradual Q_b values and constant temperatures of 22 and 24.5 °C in the water bubbler and the humidity mixer, respectively. This was to investigate the stability of U values throughout the whole water evaporation arrangement. These experiments were performed for two different Q_m values. Ultimately, the results obtained by the aforementioned computing system were plotted as shown in Figure 3.9.

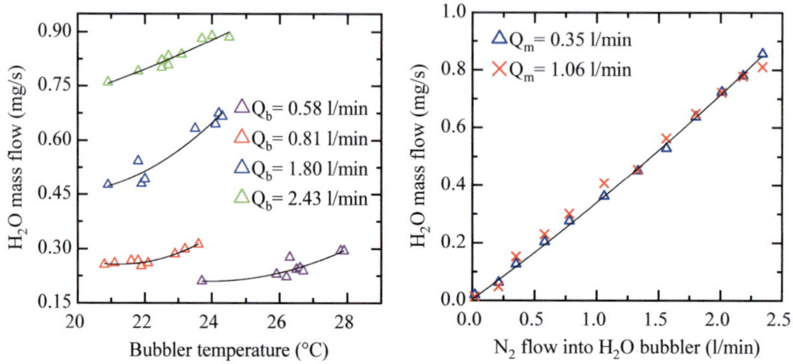

Figure 3.9: Graphs demonstrating **(left)** the H_2O mass flow (into the bubbler and the mixer) as a function of the water-bubbler temperature without a dry N_2 flow into the mixer, and **(right)** the H_2O mass flow relation with the N_2 flow rate into the humidity mixer at elevated N_2 flow rates into the water bubbler and constant T_b and T_m

At $Q_m = 0$, the AH was measured to be constant into both the water bubbler and humidity mixer at the same T_b. The H_2O mass flow was, therefore, shown to be identical at the water bubbler and the humidity mixer outlets. Figure 3.9 (left) revealed that the higher the T_b, the higher the U_b and U_m. This is interpreted by the increase of H_2O concentration within a unit volume (AH) of dry N_2. See equations (3.20) and (3.21). Consequently, the water evaporation rate increases for the same N_2 flow rate. U value was shown to increase gradually with Q_b all along the evaporation arrangement (Figure 3.9 left and right), as well. This is attributed to the increase of RH_b at higher P_w as previously proven, by which higher AH_b is induced. However, this holds true only up to $RH_b = 100\%$. At higher Q_b values than that correspond to 100% RH_b the AH_b remains at the maximum value at the investigated temperature (see clarification of Figure 3.8). Nevertheless, U continues to increase due to the increased N_2 volume flowing into the water bubble per unit time. The investigation of Q_m influence upon the U value indicated that it is constant in both the water bubbler and humidity mixer at the same T_b and Q_b, irrespective of the T_m and Q_m. See Figure 3.9 (right). This is due to the fact that no additional water vapor is introduced to the mixer by further evaporation.

3.3.5 $SiCl_4$-to-H_2O molar ratio

A complete reaction between $SiCl_4$ and H_2O molecules is essential for saving material, predicting deposition rate, and obtaining a reliable quality of the deposited oxide film (will be proven in chapter 5). For that to be achieved, care should be taken about the exact molecule concentration into the reactants' mixture. This is made possible by acquiring the suitable stoichiometry of the compounds during the reaction. Stoichiometric reactant media can be engineered with a precise calculation of the reactants' mole number using, for our case of study, equation (3.7). This is measured by what is called "molar ratio". Therefore, the mass/molecule concentration of each reactant into its relevant stream is possible to predict prior to the SiO_{2-x} APCVD system operation.

From the equation (3.7) we deduce that for each mole of $SiCl_4$ molecules two moles of H_2O molecules are required to complete the reaction and obtain one mole of SiO_{2-x} molecules, besides the four moles of the by-product HCl gas. Consequently, for fulfilling the desired stoichiometric reaction the $SiCl_4$-to-H_2O molar ratio is supposed to be 0.5. According to this value, the mass/molecule concentration of the reactants and operating boundary conditions are precisely defined using the flowing equations:

$$MR = \frac{N_{SiCl4}}{N_{H2O}} \qquad (3.27)$$

where:

MR is the $SiCl_4$-to-H_2O molar ratio.

N_{SiCl4} and N_{H2O} are the mole number of $SiCl_4$ and H_2O molecules, respectively (*mol*). This expresses the mole concentration within the investigated volume of the vapor stream.

This can be converted to mass ratio, or so called the weight ratio, as follows:

$$WR = \frac{M_{SiCl4} \, . \, N_{SiCl4}}{M_{H2O} \, . \, N_{H2O}} = \frac{M_{SiCl4}}{M_{H2O}} MR \qquad (3.28)$$

where:

WR is the $SiCl_4$-to-H_2O mass ratio.

M_{SiCl4} and M_{H2O} are the molar weight of $SiCl_4$ and H_2O molecules, respectively (g/*mole*).

Therefore, other expressions to calculate the desired mass concentration of both $SiCl_4$ and H_2O molecules flowing to the APCVD injector are:

$$K_{C(SiCl4)} = M_{SiCl4} \,.\, N_{SiCl4} \tag{3.29}$$

$$AH_m = M_{H2O} \,.\, N_{H2O} \tag{3.30}$$

The mass ratio can be, consequently, calculated using the reactant concentrations as follows:

$$WR = \frac{K_{C(SiCl4)}}{AH_m} = \frac{E_{SiCl4}}{U_m} \tag{3.31}$$

Based on the above the equation to calculate the flow rates of the reactants' carrier gas are:

$$Q_{SiC4} = \frac{E_{SiCl4}}{WR.AH_m} \tag{3.32}$$

$$Q_{(b+m)} = \frac{U_m}{K_{C(SiCl4)}} WR \tag{3.33}$$

If the oxide film deposition was restricted only upon the substrate top surface, we could estimate its thickness by the equation:

$$t_{SiO2-x} = \frac{A_S}{\rho_{SiO2-x}} 10^7 \tag{3.34}$$

Where:

t_{SiO2-x} is the thickness of the SiO_{2-x} film (*nm*).

A_s is the surface area of the substrate (*cm^2*).

ρ_{SiO2-x} is the measured density of the deposited SiO_{2-x} film (*g/cm^3*).

3.4 Advantages of the newly developed SiO_{2-x} APCVD system

The following list of advantages of this novel approach shows the benefits which outweigh the conventional methods:

1) Cost-effective deposition method indicated by the absence of vacuum pumps, catalysts and plasma generators, inexpensive chemical precursors, simple tool engineering and low-cost manufacturing materials.

2) Because the deposition process is performed at or below the room temperature, no need for additional thermal budget for heating up the mounting chuck, reactant media or reactor walls to high temperatures as in traditional APCVD, LPCVD, PECVD or ALD. Eventually, low thermal budget leads to additional drop in the overall costs.

3) Low danger of hazards due to both the $SiCl_4$ evaporation inside a tight apparatus and the release of inflammable precursors and by-products through the outward stream. Consequently, lower number of security measures and tool complexity provide safe and sufficient performance.

4) Less periodic cleaning of the APCVD apparatus than that practiced for conventional ones, thereby less effort, energy and operation interruptions, thereby adding a further drop in the overall costs.

5) Humidity mixer is a new approach, which enables a periodically precise adjustment of the absolute humidity into the relevant vapor stream. Additionally, the wick-kind vaporizer allows for suitable $SiCl_4$ evaporation rates necessitated for stable APCVD operating conditions. Consequently, maintenance of constant concentration of injected reactants is ensured, hence consistent deposition rates is guaranteed all along the APCVD system operation.

6) Thanks to the accurate control of reactant concentrations within the reactant media, wide range of the deposition rates can be obtained, according to the desired degree of stoichiometry, density, and other properties of the oxide film.

7) The new SiO_{2-x} APCVD is well-suited technique for numerous room-temperature applications such as those utilizing partially wet polymer layers, low melting-temperature metal layers, polymer substrates, microelectronics, and solar cell precursors.
8) Since the new SiO_{2-x} APCVD technique is contact free (no objects can get in contact with the substrates), metal contaminant free (because the whole setup is made of plastic), oxide particles free (no oxide dust is produced or embedded on the deposited films), and low temperature (no substrate deflection), substrates' yield is expected to be as high as 99 %.
9) Due to the fact that $SiCl_4$ and HCl vapors do not react with plastic materials, the constituting parts of the APCVD reactor are stable against corrosion and damage during operation.
10) It is feasible to handle the by-products as it is HCl gas which can be readily neutralized be means of dilution in water.
11) Virtually complete absence of material wastage since most of the reactants are deposited onto the substrate surface prior to opening the exhaust.

3.5 Summary

In the space of experimental research, an APCVD method was established for the deposition of SiO_{2-x} films, uniquely employed for solar cell applications. A novel utilization of the inexpensive substance precursors $SiCl_4$ and H_2O under atmospheric conditions was proposed for the first time in this research. This APCVD tool was constituted of polycarbonate and 3D printed thermoplastic to be adopted as a cost-effective, yet robust approach currently for laboratory exploitation. The tool comprises, most importantly, an APCVD apparatus including a sealed vapor injector and reactant-liquid vaporizers operated at constant temperatures. Purely dry N_2 gas stream was used as a carrier gas of the two chemical vapors. $SiCl_4$ vaporizer was self-built utilizing wicking principle, by which, besides its continuous evaporation, the $SiCl_4$ liquid level remains constant over the course of tool operation. Consequently, the relevant saturation vapor pressure remains constant and hence an essentially consistent $SiCl_4$ evaporation rate. H_2O evaporation arrangement was self-built comprising a DI-water bubbler attached to a humidity mixer utilizing two N_2 gas lines. The bubbler is responsible for producing 100% RH gas stream, while the mixer is installed to adjust the ultimate AH into the N_2+H_2O gas mixture prior to the introduction to the APCVD injector. It was found that a precise determination of H_2O content into the working stream was successfully possible by the adjustment of AH, or so-called H_2O molecule concentration, at the outlet of humidity mixer. Boundary conditions of the two mentioned vaporizers were investigated as to their relevant operating stability. Temperature and flow rates set for the vaporizers were found to widely influence the reactant concentrations introduced to the injector. It was concluded that the higher the temperature and flow rate into the $SiCl_4$ vaporizer, the higher the evaporation rate and the lower the $SiCl_4$ mass concentration produced. On the other hand, the higher the flow rate into the water bubbler the higher the both H_2O mass flow and AH. This conclusion holds true only at RH into the bubbler lower than 100 %. At RH of 100 % it was found that the higher the flow rate the higher the evaporation rate, yet the same the AH into the bubbler. However, the higher the bubbler temperature, the higher the H_2O mass flow and concentration into the relevant stream. It is noteworthy that the operating boundary conditions (e.g. Q and T) should be first estimated based on the calculation of molar ratio between $SiCl_4$ and H_2O molecules. When stoichiometric SiO_x film is intended, the $SiCl_4$-to-H_2O molar ratio was shown to be 0.5. Subsequently, the $SiCl_4$-to-H_2O ratio of mass concentration was 4.7.

3.6 Bibliography

[1] Krishna Seshan, Handbook of Thin-Film Deposition Processes and Techniques, William Andrew, New York, 2001.

[2] D.M. Dobkin, M.K. Zuraw, Principles of Chemical Vapor Deposition, Springer Science & Business Media, Berlin/Heidelberg, 2003

[3] K. Okuyama, D. Huang, J.H. Seinfeld, N. Tani, Y. Kousaka, Aerosol formation by rapid nucleation during the preparation of SiO_2 thin films from $SiCl_4$ and O_2 gases by CVD process, Chem. Eng. Sci. 46 (1991) 1545–1560. https://doi.org/10.1016/0009-2509(91)87004-V

[4] J.D. Chapple-Sokol, C.J. Giunta, R.G. Gordon, A Kinetics Study of the Atmospheric Pressure CVD Reaction of Silane and Nitrous Oxide, J. Electrochem. Soc. 136 (2019) 2993–3003. https://doi.org/10.1149/1.2096390.

[5] K.K. Akurati, R. Dittmann, A. Vital, U. Klotz, P. Hug, T. Graule, M. Winterer, Silica-based composite and mixed-oxide nanoparticles from atmospheric pressure flame synthesis, J. Nanoparticle Res. 8 (2006) 379–393. https://doi.org/10.1007/s11051-005-9024-y.

[6] S.K. Ignatov, P.G. Sennikov, A.G. Razuvaev, L.A. Chuprov, O. Schrems, B.S. Ault, Theoretical study of the reaction mechanism and role of water clusters in the gas-phase hydrolysis of SiCl4, J. Phys. Chem. A. 107 (2003) 8705–8713. https://doi.org/10.1021/jp034618h.

[7] N.P. Murphy, R.H. Petrmichl, Highly durable hydrophobic coatings and methods, United States Patent, 2004.

[8] Hernan Miguez, G.A. Ozin, S.M. Yang, N. And, Tetreault, Mechanical stability enhancement by pore size and connectivity control in colloidal crystals by layer-by-layer growth of oxide, Patent Application Publication, 2008.

[9] M. Sawada, K. Yamaguchi, K. Momose, H. Saka, Light Emission Life-Span Characteristics of Electroluminescent Phosphor Encapsulated by TiO_x and SiO_x Films, J. Electrochem. Soc. 148 (2001) H103. https://doi.org/10.1149/1.1387240.

[10] O.P. Prat, T. Cloitre, R.L. Aulombard, Thermal and mechanical properties of silicon tetrachloride ($SiCl_4$) and germanium tetrachloride ($GeCl_4$) in their vapor and liquid phases, Chem. Vap. Depos. 13 (2007) 199–203. https://doi.org/10.1002/cvde.200604242.

[11] National Center for Biotechnology Information. PubChem Database. Water, CID=962, https://pubchem.ncbi.nlm.nih.gov/compound/962 (accessed on June 5, 2020)

[12] P. For, A.N.E.W. Guideline, OECD Guidelines for the Testing of Chemicals, (2003) 1–13. https://doi.org/10.1787/9789264069626-en.

[13] K.S. Pitzer, D.Z. Lippmann, R.F. Curl, C.M. Huggins, D.E. Petersen, The Volumetric and Thermodynamic Properties of Fluids. II. Compressibility Factor, Vapor Pressure and Entropy of Vaporization, J. Am. Chem. Soc. 77 (1955) 3433–3440. https://doi.org/10.1021/ja01618a002.

[14] R.K. Sinnott, Chemical Engineering Design Volume 6, Elsevier, Amsterdam, 2005.

[15] Van den Bosch, C. J. H., Weterings, R. A. P. M, Methods for the calculation of physical effects – Yellow Book, CPR 14E, 3rd Ed., 2nd Rev. print, TNO, The Hague, 2005.

[16] EPA, Environmental Protection Agency, Risk management program guidance for off-site consequence analysis, publication EPA-550-B-99-009, Section D.2.3, Appendix D, Washington D.C, 1999.

[17] E.J. Kroliczek, S. Wrenn, Kimberly R., and A. Wolf, Loop heat pipe incorporating an evaporator having a wick that is liquid superheat tolerant and is resistant to back-conduction, United States Patent, 2006.

[18] Humidity conversion formulas- Calculation formulas for humidity, Vaisala Oyj, Helsinki, (2013).

[19] Air - Water Vapor Mixtures, https://www.ohio.edu/mechanical/thermo/Applied/Chapt.7_11/Chapter10b.html

4 APCVD injector architecture and CFD study

4.1 Introduction

Uniform thickness and uniform stoichiometry are the two main prerequisites to be considered when it comes to chemical vapor deposition (CVD) of thin solid films. The homogeneous suspension of the reactive species distributed equally all over the substrate surface leads to consistent films with a satisfactory thickness uniformity. This homogeneous mixture is realized by carefully tuning the mean free path length of the reactant ions/molecules and the activation energy necessary for boosting the chemical reaction rate. Besides, the reaction time is another aspect to be considered, which assures the chemical stoichiometry, material purity, and uniform distribution in the deposited film. Larger mean free path lengths of the reactant ions leads to a more uniform reactant mixture, more stoichiometric product, and thicker film. However, at atmospheric pressure the mean free path length of the reactant species is generally lower than 1 µm and hence this parameter is not suited for improving the film properties and the uniformity of the deposition. On the other hand, a higher reaction activation-energy leads to greater mobility and collision frequency of the ions/molecules, and hence a higher concentration of dissociated reactant species. These causes and effects stimulate a complete reaction between the reactants, hence high deposition rates and consistently composed products without impurities.

In conventional CVD techniques, especially those employed to prepare SiO_2 films, the activation energy for the reaction is provided by setting high deposition temperatures in the range of 400 – 2300 ºC [1,2][3]. For relatively low-temperature deposition systems, in the range of 25 – 150 °C, dissociating the reactant precursors is promoted either by plasma assistance adopted in plasma-enhanced CVD (PECVD) and few atmospheric pressure CVD (APCVD) systems [4–7] or by reaction catalysis utilized in e.g. atomic layer deposition (ALD) [8,9]. The efficient ion mobility, due to the bigger mean path way, is usually promoted by adjusting ultra-low chamber pressure (< 1 millitorrs) as for ALD, low-pressure CVD (LPCVD), and numerous PECVD techniques [8,10,11].

Currently, there are two ways of introducing and intermixing the gas precursors into the reaction chambers, depending on the used CVD technique. In PECVD and LPCVD tools, two or more gas streams are driven simultaneously to the chamber, whereby the flowing gases disperse adequately with one another because of the low pressure and relatively large chamber volume, which enables larger mean path of the reactant species [12–14]. As for APCVD tools, two or more reactive vapors carried by an inert gas (Ar or N_2) are introduced into an injector positioned above the substrates [15]. Intermixing of the reactive gases occurs directly above the substrate surface and the reactant product is deposited in situ, creating the desired film. The reaction by-products are out-streamed through the associated exhaust chamber, facilitating continuous recharging of the reaction chamber with fresh gases. The main drawback of LPCVD and PECVD techniques, besides their structural complexity, is the reduced reactant concentrations within the flowing media, which induces inconsistent deposition rate within the chamber. Moreover, in the case of hot-wall reactors, there is a higher probability of forming parasitic dust, which agglomerates from the gas precursors and condenses onto the chamber inner walls. Consequently, the dust agglomerants could fall on the substrates causing the deterioration of the deposited film properties, besides the high wastage of the gas precursors. Similarly, the APCVD injectors cause parasitic dust formation as a consequence of the abrupt gas-phase reaction between the molecules and the immediate conversion from intermediate to ultimate products (due to low mean pathway under atmospheric pressure). The occurrence of the gas-phase reaction far from the substrate surface and the exhaust location close to the injector outlet-openings lead to much more vapor precursor wastage than that of the LPCVD and PECVD techniques. Numerous approaches to overcome the effects of the substrate's geometries are summarized in the next section.

In the newly developed APCVD method for the preparation of silicon oxide films utilizing $SiCl_4$ and H_2O vapors, there are two significant challenges to obtaining chemically homogeneous mixture and uniform films. Firstly, the deposition under atmospheric conditions (1 atm and 25 °C or below) induces the following: i) the mean pathway of the reactant molecules is much shorter than that of the evacuated techniques, and ii) the energy of the reaction initiation and completion is much lower than that of thermal deposition techniques. Secondly, the abrupt nature of the reaction between $SiCl_4$ and H_2O (occurs in few milliseconds) causes instant oxide dust formation on the injector inner walls, which may contaminate the substrate surface. These effects obstruct the possibility to intermix the two reactants at the substrate surface and realize the uniform distribution of the mixture over a wide substrate area. These mentioned obstacles give rise to the development of a new APCVD injector/apparatus that enables efficient interaction between the reactant molecules, thus resulting in a complete reaction without wall contamination with the oxide dust. Furthermore, a reasonable deposition thickness-uniformity over 6-inch wide substrates is realized.

In this chapter, a novel APCVD injector made of three-dimensional (3D)-printed thermoplastic is introduced for the preparation of SiO_{2-x} films at room temperature utilizing vaporized $SiCl_4$ and H_2O precursors. Vigorous mixing of the two reactants before striking the substrate surface and uniform distribution of the gas mixture over a relatively large substrate area are attempted. A computational fluid dynamics (CFD) study of the gases flowing into the APCVD injector portions is performed after an introduction about the fluid mechanic concepts and equations needed for this study. At first glance premixing the H_2O and $SiCl_4$ vapors might seem to induce a gas-phase reaction and hence film deficiency. A solution to minimize the impact of abrupt gas-phase reaction within the apparatus volume is investigated. For this purpose, a trade-off relation is adjusted between the precursors/product concentration, pressure, exhaust flow, and film deposition rate. A novel solution to eliminate the parasitic deposition of oxide particles onto the reactor inner walls is briefly introduced, as well. Suppression of the presence of one reactant upon the reactor inner walls is effectively investigated to prevent the heterogeneous reaction with unintended surfaces.

4.2 A brief review of APCVD injectors

The most common APCVD injectors for preparing silicon oxide films utilizing different chemical substances are demonstrated in [16–18], where different approaches are adopted to overcome the homogeneous reaction influence. In [16] a linear-type injector of TEOS and ozone is employed to introduce individual reactants to the substrate surface via tightly spaced outlet slits. As a result, two spaced streams of various gas precursors are obtained. The gas streams feature a high flow velocity, by which the reactant mixing is avoided before reaching the substrate surface. Consequently, the diffusion of the reactants across the recirculation regions on either side of the injector slits occurs in a time longer than that of the residence time of the freely flowing gases. Thanks to the linear motion of the substrate, the other reactant diffuses and then reacts with the previously adhered one upon the substrate surface. Subsequently, the desired film is deposited, and the by-products are desorbed and driven to the exhaust together with non-reacted precursors.

The approach used in the industrial APCVD injector from Schmid group [17] is to inject various gas precursors separately through multiple inlets so as to react after passing the injector outlet openings. The substrates are mounted few millimeters below the injector openings, thus the reaction between the precursors takes place as close as possible to the substrate surface. In addition, curtain flows of an inert gas such as N_2 are used to segregate the various gas streams in the small gap between the injector outlets and the substrates surface. Therefore, no molecular-diffusion takes place across the recirculation regions. The lateral exhaust is responsible for the immediate release of the by-products so as to prevent forming undesired particles/contaminants. In reference [18], a premix of TEOS and O_2 containing lower concentration of O_3 (0.1 – 1 %) is prepared in a mixer located away from the respective APCVD injector so as to catalyze the subsequent reaction. The lower-concentrated O_3/TEOS mixture is then driven to the injector to be mixed vigorously at the outlet with another gas stream containing higher concentration of O_3 (1 – 10 %). The injector comprises series of slit-type channels that convey the two aforementioned streams as close as possible to the substrate surface. However, this approach is insufficient for completely impeding the impact of the gas-phase reaction and parasitic deposition. This is due to the fact that the low O_3 concentration, contained in the O_2 stream, premixes with TEOS in the delivering pipes leading to a light oxide deposition in these pipes and in the injector nozzles. There are other chemicals alternatively utilized to deposit the SiO_2 films using this injector design. An organo-silicate compound vapor is premixed mildly with an extremely low concentrated vapor comprising H_2O and H_2O_2, apart from the injector. The light mixture is then injected through the linear injector to actively get in contact with the third vapor (triethoxyfluorosilane, silicon tetrafluoride, or silicon tetrachloride) driven from another injector. Subsequently, the SiO_2 film is deposited on the substrate surface.

4.3 Fundamentals of fluid dynamics

The study of fluid dynamics is based on three fundamental laws, the conservations of mass, momentum and energy. In the study of the APCVD injector the gases are considered incompressible (as they are in continuous motion and feature a constant density), as well as, their flow is inviscid and adiabatic. For that matter, theses conservation laws are described by Euler equations [19] which can be considered as a particular expression of Navier-Stokes equations with viscosity and thermal conductivity equal to zero. The derivations given in the following sections are obtained by applying the conservation equations to a fixed volume in space, also known as the control volume (V), where the corresponding surface boundary is called the control surface (A).

4.3.1 Law of Conservation of Mass and the Continuity Equation

This law considers that the mass of a fluid passing a fixed volume in space (the flow rate of the mass flow) is unchangeable. Mathematically, this translates to the statement that the derivation of the fluid mass with respect to time is zero, expressed by [19]:

$$\frac{dm}{dt} = 0 \tag{4.1}$$

where (m) is the fluid mass and (t) is the time.

This is also called the continuity equation for the mass of a control volume. This can be stated in an integral form as follows [20]:

$$\frac{d}{dt}\int_V \rho\, dV = \int_V \frac{\partial \rho}{\partial t} dV \tag{4.2}$$

where (ρ) is the material density, and (V) is the fixed volume.

It is to be noted that in this equation the time derivative is taken inside the right-side integral due to the fact that in a control volume the surface boundary A is fixed and thus its velocity (u_A) is equal to zero. Therefore, the integration limit (the control volume V) is independent of time.

On the other hand, the rate of fluid mass flowing outwards (denoted by minus) of the fixed surface surrounding a control volume is the surface integral given by [20]:

$$-\int_A \rho\, u\, \vec{n}\, .dA = -\int_A \rho\, u \cdot dA \tag{4.3}$$

where (u) is the velocity vector of fluid mass crossing the surface boundary (A) and (n) is the unit outward-pointed normal vector to the surface boundary (the vector is perpendicular to the surface), which is equal to one.

Complying with the mass conservation law, we can deduce the following:

$$\int_V \frac{\partial \rho}{\partial t} dV = -\int_A \rho\, u \cdot dA \tag{4.4}$$

Therefore

$$\int_V \frac{\partial \rho}{\partial t} dV + \int_A \rho\, u \cdot dA = 0 \tag{4.5}$$

The surface integral of the mass flow rate can be transformed to a volume integral by means of applying the divergence theorem and vector calculus applied for three-dimensional flow. The equation thus becomes:

$$\int_V [\frac{\partial \rho}{\partial t} dV + \nabla \cdot (\rho\, u)]\, dV = 0 \tag{4.6}$$

When the flow is steady, the partial derivative with respect to time equals zero. Therefore, the final continuity equation becomes:

$$\int_V \nabla \cdot (\rho\, u)\, dV = 0$$

$$\nabla \cdot (\rho\, u) = 0 \tag{4.7}$$

Since ∇ is a vector operator, this equation can be expressed according to Cartesian Coordinates as follows:

$$\frac{\partial}{\partial x}(\rho\, u_x)\vec{\imath} + \frac{\partial}{\partial y}(\rho\, u_y)\vec{\jmath} + \frac{\partial}{\partial z}(\rho\, u_z)\vec{k} = 0 \tag{4.8}$$

If the steady flow is studied in one-dimension the mass continuity equation can be expressed as follows [19]:

$$\sum \dot{m} = \sum \rho\, u\, A = 0 \tag{4.9}$$

where $(\dot{m})$ is the mass flow rate of the fluid (*kg/s*).

When the flowing fluid is incompressible (ρ = constant), the mass continuity equation simplifies to the volume continuity equation:

$$\nabla \cdot u = 0$$

$$\sum \dot{V} = \sum u \cdot A = 0 \tag{4.10}$$

Where $(\dot{V})$ is the volume flow rate of the fluid (m^3/s).

4.3.2 Law of Conservation of Momentum

The second governing law to describe the fluid dynamics is the law of conservation of momentum, which is derived from the Newton's second law of motion. This law states that the change in the momentum of an object, e.g. a fluid quantity in a control volume, during time is proportional to the applied net force, which is expressed for linear cases by the vector equation [19]:

$$\sum F_{ext} = \frac{1}{g_c} \frac{d\overrightarrow{M}}{dt} \tag{4.11}$$

where (M) is the momentum of fluid inside a control volume, (ΣF_{ext}) is the summation of all external forces acting on the fluid in the control volume, and (g_c) is the proportionality factor which is given as:

$$g_c = 1 \frac{Kg\,.\,m}{N \cdot s^2}$$

The external forces acting on a fluid mass that is present within a control volume are gravitational forces (mass), surface forces (pressure, friction/viscosity), and reaction/anchoring forces, also known as molecular momentum transport-forces [20,21]. The equation of linear momentum conservation is derived by means of applying the Reynolds transform theorem [19] to the equation (4.11). Accordingly, the transformation of this equation yields in:

$$\sum \overrightarrow{F_{ext,V}} = \frac{1}{g_c}\frac{\partial}{\partial t}\int_V \rho\,\vec{u}\,dv + \frac{1}{g_c}\int_A \rho\,\vec{u}\,(\vec{u}\cdot\vec{n})\,dA \qquad (4.12)$$

In this vector equation it is enough to consider one component, e.g. the *x* one, thus it simplifies to:

$$\sum F_{x,ext} = \frac{1}{g_c}\frac{\partial}{\partial t}\int_V \rho\,u_x\,dv + \frac{1}{g_c}\int_A \rho\,u_x\,(u_x\cdot\vec{n})\,dA \qquad (4.13)$$

Under steady state conditions the first integral in this equation is equal to zero. Incompressible flow in one dimension normal to the surface (A) leads to constant values of (ρ) and (u) across any studied cross section of the flow. Therefore, the equation becomes:

$$\sum F_{ext} = \rho\,u\cdot u\int_A dA \qquad (4.14)$$

Solving the integral equation leads to the linear momentum equation:

$$\sum F_{ext} = \sum \dot{m}\cdot u \qquad (4.15)$$

4.3.3 Law of Conservation of Energy

It is necessary to evaluate the energy required for a process to take place or the energy harvested from a device during operation. By way of example, the energy required for the $SiCl_4/H_2O$ gas mixture to flow and distribute all across the APCVD injector can be approximately estimated by applying the law of conservation of energy. The conservation of energy for a control volume is a statement of the first law of thermodynamics for an open system. It states that the change of a system's energy with respect to time is equal to both heat exchange (due to temperature differences) and work exchange (due to boundary movement and flow of mass through the control volume) between the system and its surrounding. This can be expressed by the following equation [19]:

$$\frac{\delta Q}{dt} = \frac{\delta W}{dt} + \frac{dE}{dt} \tag{4.16}$$

where:

$\delta Q/dt$ and $\delta W/dt$ are the rates of heat and work transfer into the open system (or to the control volume) at any instant of time, respectively.

dE/dt is the system's energy derivative.

Applying the Reynolds transform theorem [19] to the system's energy derivative leads to the equation:

$$\frac{dE}{dt} = \frac{\partial}{\partial t}\int_V \rho\, e\, dv + \int_A \rho\, e\,(\vec{u}\cdot\vec{n})\, dA \tag{4.17}$$

where (E) is the system's total energy and (e) is the specific energy (the total energy per unit mass). The latter can be expressed, considering that the total energy of an open system is the summation of internal, kinetics (work) and gravitational/potential energies, as follows [19]:

$$e = \tau + \frac{1}{2}u^2 + gz$$

where (τ) is the specific internal energy, (u) is the flow velocity, (g) is the gravitational acceleration, and (z) is the elevation.

For steady flow the equation (4.17) becomes:

$$\frac{\delta Q}{dt} = \frac{\delta W}{dt} + \int_A \rho\,(\tau + \frac{1}{2}u^2 + gz)\,(\vec{u}\cdot\vec{n})\, dA \tag{4.18}$$

For one-dimensional flow (e, ρ, and u) are constant over any cross section and (u) is normal to the surface (A). Therefore, the energy equation for a control volume is shortened to:

$$\frac{\delta Q}{dt} = \frac{\delta W}{dt} + \sum \dot{m} \left(\tau + \frac{1}{2} u^2 + gz \right) \quad (4.19)$$

4.3.4 Bernoulli principle

This principle expresses the inverse relationship between the pressure, and the velocity and elevation of any point of a single fluid flow. It states that as the fluid flows in a certain direction and a change has occurred either in its velocity between two regions, e.g. due to change in its cross-sectional diameter, or in its elevation, an inverse change in its relevant pressure will occur simultaneously between the two inspected regions. The pressure explained in this principle is the static pressure of the fluid which is an internal pressure exerted by the neighboring fluid particles on one another, as well as, by the particles at the outermost flow in the outward direction applied on the surrounding walls (when the flow is inside a duct).

Bernoulli equation is important to study fluid flows that experience changes in their velocity (kinetic energy), pressure (external work) or elevation (gravitational potential energy). It can be derived from Euler equation or the conservation equations for inviscid flows [19–21]. Bernoulli equation can be described by the derivative form:

$$\frac{d}{ds}\left(\frac{1}{2} u^2\right) + \frac{1}{\rho}\frac{dp}{ds} + g\,\frac{dz}{ds} = 0 \quad (4.20)$$

Where (d/ds) is the distance derivative along a streamline, (dp/ds) is the pressure derivative along the streamline, (g) is the gravitational acceleration, and (dz/ds) is the elevation derivative along the streamline.

Integrating of the equation along the streamline leads to:

$$\frac{1}{2}\sum u^2 + \int \frac{1}{\rho} dp + g \sum z = 0 \quad (4.21)$$

which simplifies to:

$$\frac{p}{\rho} + \frac{1}{2}u^2 + g\,z = constant \tag{4.22}$$

4.3.5 Reynolds number

Reynolds number is a nondimensional parameter used for characterizing the flow pattern of Newtonian and incompressible fluids [21]. It fundamentally describes the ratio of inertial forces to viscous forces for such fluids. The inertial forces arise between fluid particles experiencing both nonuniformly distributed pressure and in-between molecular friction inducing fluid movement. This can be evaluated by the acceleration of a fluid particle given by [22]:

$$|(u \cdot \nabla)\, u| \approx \frac{u^2}{L}$$

whereas, the viscos forces arise from the cohesion and attractive forces between adjacent fluid particles resisting their relevant flow. the value of the viscous force is evaluated by:

$$|\nu\, \nabla^2\, u| \approx \frac{\vartheta\, u}{L^2}$$

Therefore, Reynolds number is calculated in a linear form by the equation [20–22]:

$$Re = \frac{u\,L}{\nu} = \frac{\rho\,u\,L}{\mu} \tag{4.23}$$

where (ρ) is the fluid density, (u) is the fluid velocity-magnitude, (L) is the characteristic dimension of the studied field, (υ) is the viscosity coefficient (or kinematic viscosity) of the fluid, and (μ) is the dynamic or absolute viscosity of the fluid.

Incompressible flows tend to change their characteristics according to Re change as follows: i) when Re is less than 2000 – 2300, the flow is characterized by laminar, ii) when Re is between 2000 and 4000, the flow is considered transitional, and iii) when Re increases more than 4000, the flow is considered turbulent. In some references the Re $\geq$ 2300 refers to turbulent flow and the transitional flow is omitted. It is worth noting that when Re $\gg$ 1 the fluid viscosity can be neglected, whereas the acceleration term of a fluid particle can be ignored when $Re \ll 1$.

4.3.6 Laminar and turbulent flows

Laminar flow is by definition a smooth flow wherein the fluid layers flow straightly parallel to each other without macroscopic mixing and it occurs at low velocities (or low Re). Turbulent flow is, on the other hand, a flow wherein irregular motions and spread fluid streams are present developing macroscopic mixing motions perpendicularly to the flow direction. It occurs at flow velocities (or Re values) way much higher than those of the laminar flow [20,21].

Both the two flow manners should be studied when the flow is fully developed. It is to consider that when the flow is fully developed, at which the velocity is represented as e.g. the *x*-directed instantaneous velocity, the flow is defined as follows:

$$\frac{du}{dx} = 0$$

This, in turn, indicates that the fully developed phase is reached when the velocity profile no longer changes along the flow path.

If we assume a laminar steady flow of an incompressible Newtonian fluid through a cylindrical duct, the velocity profile of the fully developed flow is a parabolic as shown in Figure 4.1. The flow velocity at any cross section is given by the equation [19]:

$$u = U_{max}\left[1 - (\frac{r}{r_0})^2\right] \quad (4.24)$$

where (U_{max}) is the mean velocity occurring at the centerline of the flow, which is the largest among the whole cross-sectional flow. (r_0 and r) are the radius of the duct cross section and the radius of the cross-sectional flow at the point of interest.

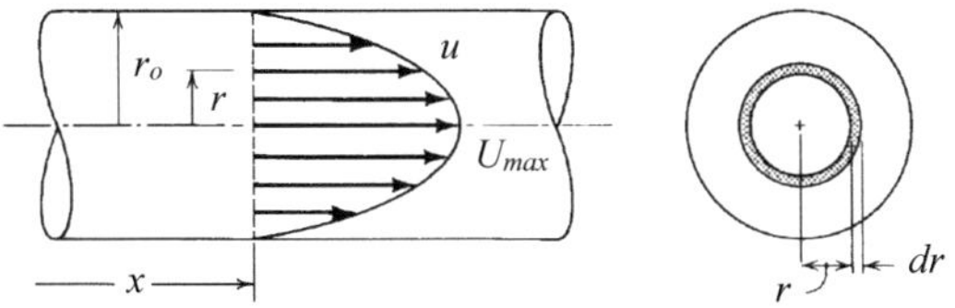

Figure. 4.1: A schematic illustration of a cross section of a cylindrical duct showing the velocity profile of the fully developed laminar flow. After [19].

If we consider a turbulent steady flow of an incompressible Newtonian fluid through the cylindrical duct, the velocity profile of the fully developed flow is relatively flat as shown in Figure 4.2. It is worth noting that in this flow manner the fluid particles present in one section feature nearly the same velocity, streamwise distribution and turbulence intensity, which closely align one-dimensional flow. The flow velocity is given by the equation [23]:

$$u = U_{max}\left[1 - \left(\frac{r}{r_0}\right)^{1/7}\right] \qquad (4.25)$$

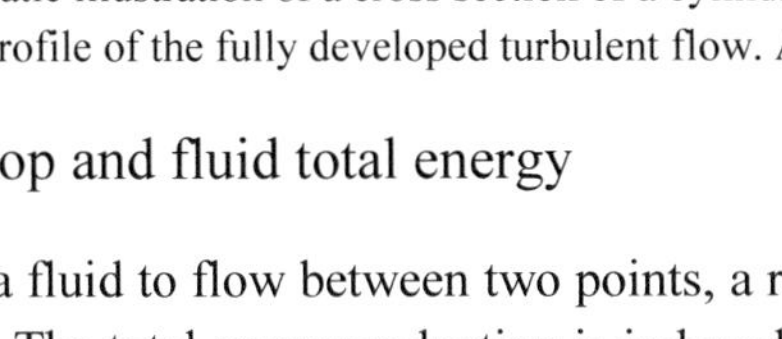

Figure. 4.2: A schematic illustration of a cross section of a cylindrical duct showing the velocity profile of the fully developed turbulent flow. After [19].

4.3.7 Pressure drop and fluid total energy

In order for a fluid to flow between two points, a reduction in its total energy must occur. The total energy reduction is induced because of friction forces and change in the fluid momentum. This corresponds with the fact that fluids flow from high-total-energy points to low-total-energy points. Back to Bernoulli's equation of the fluids characterized by an adiabatic system and flowing between two points, the total energy drop, or so-called head loss, should be added to the associated energy of the fluid (the summation of pressure, kinetic and potential energies). The head loss of a flow through a pipeline is a function of numerous parameters: the inside diameter of the pipe, its length, the roughness elements of the pipe inner walls, the average flow velocity, and the fluid density and viscosity [20]. The head loss (H_L)is given by the so-called Darcy-Weisbach equation [21] as follows:

$$H_L = f \, \frac{L \, u^2}{D \, 2g} \tag{4.26}$$

with (f) being the Darcy-Weisbach friction factor, (L) the flow distance or the pipe length, (D) the pipe inside diameter, (u) is the average flow velocity (obtained from the continuity equation), and (g) the gravitational acceleration. (f) is approximated by Haaland equation [21] for cylindrical duct shape as follows:

$$f = \left[-0.782 \; ln\left\{\frac{6.9}{Re} + \left(\frac{\varepsilon}{3.7D}\right)^{1.11}\right\}\right]^{-2} \tag{4.27}$$

with (ε) being the roughness of the pipe material and (Re) is Reynolds number of the flow.

The modified Bernoulli's equation between two points becomes:

$$\frac{P_1}{\rho} + \frac{1}{2}\,u_1^2 + g\,z_1 = \frac{P_2}{\rho} + \frac{1}{2}\,u_2^2 + g\,z_2 + f\,\frac{L\,u^2}{D\,2g} \tag{4.28}$$

Consequently, it can be deduced that the static pressure drop of a fluid flow occurs because of three components: increase in flow velocity (or flow rate), increase in elevation, and, most importantly, head losses, e.g. the friction loss which is subsequently translated to heat loss.

It is to conclude that the head loss reduces the total static pressure of a flowing fluid irrespective of the flow velocity and elevation.

4.4 Newly developed SiO_{2-x} APCVD injector arrangement

A newly-developed SiO_{2-x} APCVD injector was manufactured using 3D printing technology, also known as fused deposition modeling. Since it operates at room temperature, it was made of polycarbonate and thermoplastic materials such as polylactic acid and acrylonitrile-butadiene-styrene. The injector was designed to be positioned vertically over the substrates, by which the gases flow from the top inlet-cap through two openings, each for one gas precursor. The SiO_{2-x} APCVD injector layout, shown in Figure 4.3, comprises four compact components connected as follows:

1) A Kenics-type static gas mixer attached to an inlet cap on top.
2) Venturi nozzle attached to the lower opening of the static mixer.
3) A hyperbolic or divergent distributor attached to the lower opening of the Venturi nozzle.
4) A deposition chuck, upon which the substrates are mounted and the later three-component arrangement is placed nearly tightly. See more details in section 4.4.3.

Detailed information about this design are available elsewhere.

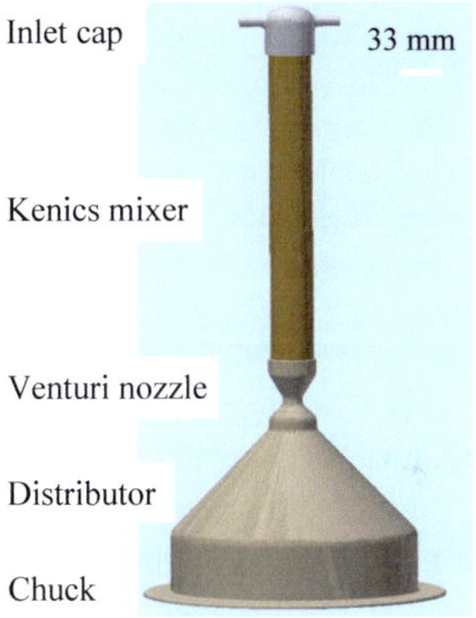

Figure 4.3: A schematic drawing of the SiO_{2-x} APCVD injector layout showing inlet openings, Kenics static mixer, Venturi nozzle, and gas distributor positioned on top of a deposition chuck

4.4.1 Investigation of the static mixer

A Kenics-type static mixer was employed to intermix the two gas streams H_2O and $SiCl_4$, which are characterized by a laminar flow. As illustrated in Figure 4.4 the mixer layout consists of 8 segments inserted into a transparent polycarbonate-plastic tube with an inner diameter of 33 mm. The successive segments are twisted opposing to one another; 180° clockwise and 180°

counter-clockwise. The segments are arranged, as such they are rotated by 90 °to one another. The length of each segment is 1.5 times its width. In the current design the dimensions are set to 32 × 48 mm. The chosen segment number and twist angle are justified later in this section.

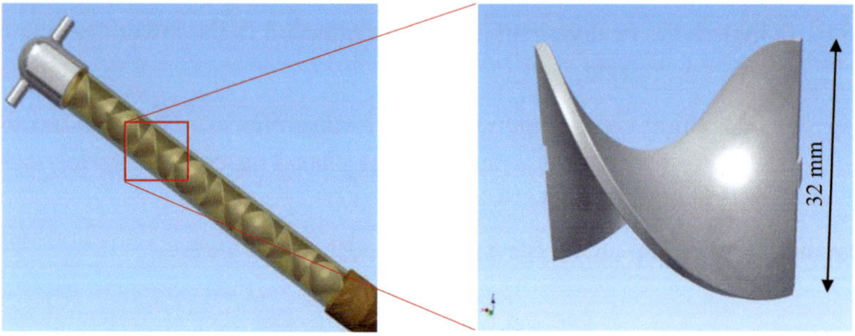

Figure 4.4: Left: A schematic picture for the Kenics-static mixer showing an assembly of 8 opposed segments inserted tightly into a tube attached to an inlet cap at the top and to Venturi nozzle at the bottom. **Right:** 180° twisted plate with dimensions of 32 × 48 mm forming a single segment.

The static mixer enables sufficient mixing of two fluids at the outlet by motionless portion assembly simulating the dynamic blending of fluids. The mixing mechanism is explained in detail in [24,25] for laminar flows. For that, two concepts are used, flow division and radial mixing. The flow division starts at the front edge of the first segment splitting the flow into two streams. At each succeeding segment, the streams are further split, resulting in an exponential growth of the flow layering as illustrated in Figure 4.5 [25]. The number of the resulting channels can be readily calculated as ($N = 2^n$) with n being the designated number of the segments. In the radial mixing each divided stream flows radially towards the tube walls and then flows back to the tube center. This induces an alternatively rotational motion of the divided streams as a consequence of the alternative difference of the twisted segment-direction.

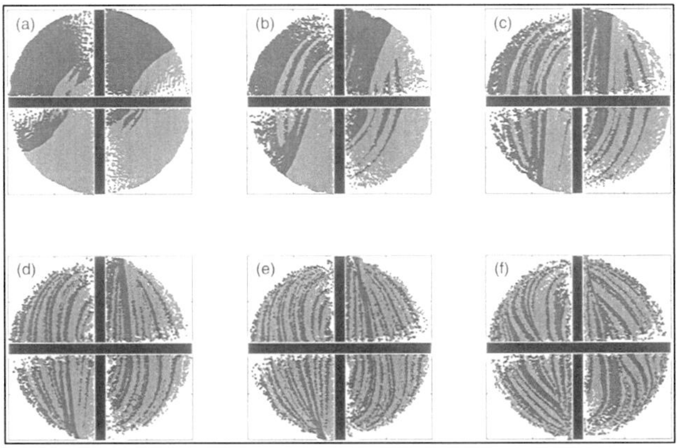

(1)

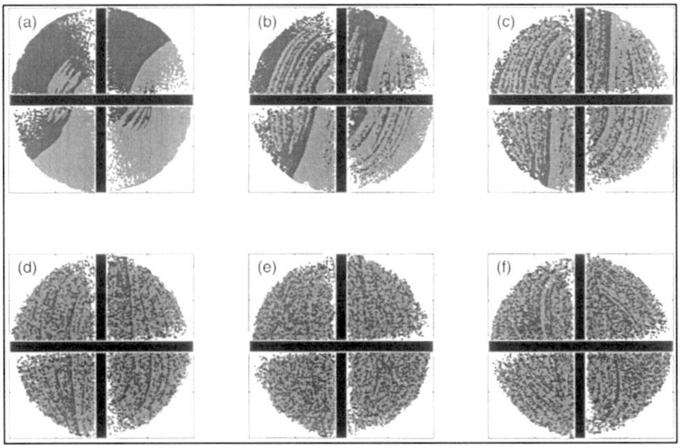

(2)

Figure 4.5: Cross-sectional profiles simulated for two initially segregated components characterized by identical volume after **(1):** 6 mixer segments and **(2):** 10 mixer segments at 6 twist angles – (a) 30°, (b) 60°, (c) 90°, (d) 120°, (e) 150°, and (f) 210°. A uniform mixture is evident after 10 segments at the twist angles 120 – 210° represented by (d) – (f) sections, respectively. From [25].

The advantage of using such mixer is that it eliminates radial gradients of the temperature, flow velocity and substance compositions. The energy required for the mixing is obtained from the pressure loss in every element as the fluids flow through the static mixer. If we ignore the energy gain from the gravitational head, the head loss in equation (4.28) possesses a considerable value, which is enough to obtain a homogeneous mixture with maintaining a high momentum transfer, hence the ability to flow. It is proven that the mixing quality is a trade-off between segment number and twist angle [24,25]. It is, for instance, shown that after 60 segments twisted at angles between 30 and 90°, the mixture is inhomogeneous as well as after 6 segments twisted at a bigger angle of 210°, seen in Figure 4.5 (1). On the other hand, after 10 segments twisted at angles between 120 and 210°, the mixture is significantly homogeneous, seen in Figure 4.5 (2). That is to say, the bigger the twist angle of the mixer segment, the less the segment number required to obtain a sufficiently homogeneous mixture, yet a significant segment number is required. Accordingly, it was suggested to use 8 mixer segments twisted at 180° for the Kenics mixer utilized in the newly-developed APCVD injector.

A test was performed to assess the quality of mixing into the self-manufactured and vertically positioned static mixer using two different colored liquids. Two beakers with blue- and yellow-colored water were poured at once into the mixer. The test was captured by a high-resolution camera. As shown in Figure 4.6, the blue and yellow water volumes are dispersed progressively along the mixer so that they exit its outlet sufficiently mixed, which is indicated by the resulting green liquid. Consequently, the design of the self-manufactured Kenics mixer was found to be robust.

Figure 4.6: Photos taken continuously when blue and yellow-colored water volumes were simultaneously poured into a self-made static mixer of the Kenics type featuring an inner diameter of 33 mm. A uniform mixture is obtained, which is indicated by the green-colored water at the outlet. The photos from left to right show the mixing progress from the start to the end, respectively.

The static mixer was, ultimately, taken into operation in the APCVD setup so as to assess its mixing performance when depositing SiO_{2-x} films. The samples to be coated were 4-inch wide as-sawn and mechanically polished Si wafers. The N_2-carried gas precursors $SiCl_4$ and H_2O flowed into the mixer at 1.5 l/min, each. The deposited films were compared with APCVD SiO_{2-x} films prepared without efficient mixing – readily performed by loading the two gas precursors via two inlets to a tube of 33 mm in diameter, without inner twisted segments. Figure 4.7 (left) demonstrates a non-uniform SiO_{2-x} film after using the simple tube. This is attributed to the laminar flow of the two gases through the tube, wherein each gas stream flows consistently on one side and the interaction takes place only in the originated stream interface. The thicker layer on the Si-wafer upper-right side is attributed to the reaction taken place at the H_2O injected side, whereas the absence of H_2O vapor at the bottom left side of the Si wafer resulted in an extremely low deposition rate. The thickness of the deposited film is low in comparison with the volume of injected precursors. The static mixer, on the other hand, enabled obtaining a significantly thick film as shown in Figure 4.7 (right).

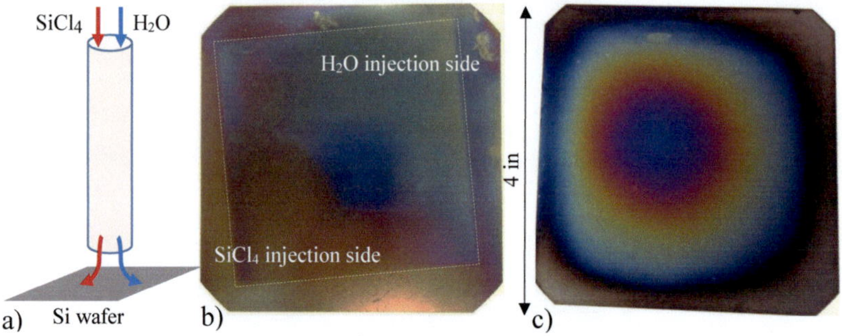

Figure 4.7: a) A sketch illustrating the insufficient mixing of $SiCl_4$ and H_2O gases during flowing through a simple tube without twisted plates. On the right side, photos of SiO_{2-x} films deposited onto 4-in^2 wide substrates placed 20 cm under the outlet of two different tubes: **b)** without inserted segments (as illustrated in Figure a) showing a non-uniform film with low thickness upon an as-sawn Si wafer, owing to insufficiently mixed gas precursors, and **c)** with inserted segments showing a thicker SiO_{2-x} film and improved layout upon a mechanically polished Si wafer, thanks to well-mixed precursors at the mixer outlet. The different contrast between b) and c) is attributed to the different surface finishing of the used Si wafers. The square imprint (marked in yellow dashed line) on the wafer seen in Figure b) is because of a paper sheet (with an area smaller than the wafer's one) used to protect the wafer surface from any contamination during storage, which was removed before the deposition.

Despite the film being non-uniform, it is symmetric and fairly thick, particularly in the center, which is not seen in the film deposited by the simple tube. Therefore, this design of the static mixer was proven to homogeneously mix the $SiCl_4$ and H_2O vapors at its outlet. The thick center and circular outline of the deposited film are attributed to extrusion of the mixture to an area as narrow as the mixer outlet-area. Moreover, since the flow at the mixer's outlet is still laminar as it features a low Reynolds' number, the boundary layer at the inner walls is slower than the rest medium. See Figure 4.1. Consequently, the stream of the mixture expands insignificantly to the sides, all along the 20-cm distance between the mixer's outlet and the substrate. Furthermore, the central flow possesses a velocity higher than that of the outermost flow, thus it reaches the substrate surface faster. These impacts lead to inconsistent distribution of the gas mixture over an area wider than that of the mixer cross-section, hence non-uniform films. These findings necessitate the creating of turbulences and vortices in the gas mixture before striking the substrate surface, by which the mixture distributes equally over a wider area promoting a uniform film thickness. A solution to obtain such a characteristic mixture was to integrate a Venturi nozzle in the injector arrangement. The relevant design and performance are explained in the next section.

4.4.2 Numerical study of the Venturi nozzle

Since moving parts are unfavorable in the newly developed APCVD injector, the way to create turbulences above the substrate surface was to cause a deflection in the gas flow when injected from the static mixer [25]. This was made possible according to the flow characteristics obtained with Venturi nozzles at low Re [26]. CFD simulations were performed with ANSYS software to investigate the dimensions, at which the Venturi nozzle develops a flow deflection at the outlet for given flow boundary-conditions. The turbulences were calculated by using the two standard $k - \varepsilon$ equations model [26,27]:

$$\frac{\partial(\rho k)}{\partial t} + \frac{\partial(\rho k u_i)}{\partial x} + \frac{\partial(\rho k u_j)}{\partial y} + \frac{\partial(\rho k u_k)}{\partial z} = \frac{\mu_t}{\sigma_k}\left(\frac{\partial^2 k}{\partial x^2} + \frac{\partial^2 k}{\partial y^2} + \frac{\partial^2 k}{\partial z^2}\right) + 2\mu_t . S_{ij} . S_{ij} - \rho\varepsilon \tag{4.29}$$

$$\frac{\partial(\rho\varepsilon)}{\partial t} + \frac{\partial(\rho\varepsilon u_i)}{\partial x} + \frac{\partial(\rho\varepsilon u_j)}{\partial y} + \frac{\partial(\rho\varepsilon u_k)}{\partial z} = \frac{\mu_t}{\sigma_\varepsilon}\left(\frac{\partial^2 \varepsilon}{\partial x^2} + \frac{\partial^2 \varepsilon}{\partial y^2} + \frac{\partial^2 \varepsilon}{\partial z^2}\right) + C_{1\varepsilon}\frac{\varepsilon}{k} - \rho\varepsilon 2\mu_t . S_{ij} . S_{ij} - C_{2\varepsilon}\frac{\varepsilon^2}{k} \tag{4.30}$$

with (k) being the turbulent kinetic energy, (ε) the dissipation of the turbulent kinetic energy, and Sij is the rate of deformation of the flow.

The constants are as follows: $\sigma_k = 1.00$, $\sigma_\varepsilon = 1.30$, $C_{1\varepsilon} = 1.44$, and $C_{2\varepsilon} = 1.92$. (μ_t) is the eddy viscosity, given by the equation:

$$\mu_t = \rho \, C_\mu \, \frac{k^2}{\varepsilon} \tag{4.31}$$

where the constant $C_\mu = 0.09$.

The way Venturi nozzle was designed is to convert the flow from laminar to turbulent at the outlet by means of careful determination of its dimensions. Since it is extensively studied in the state-of-the-art literature, variation of nozzle dimensions was excluded from investigating in this research, and instead, ultimate designs were adopted as to the desired nozzle function. Based on the prior art, the nozzle dimensions were proposed, as though the contraction ratio and the diffusion angle are as small as possible. This, consequently, results in asymmetric distribution of the stream velocities in the divergent portion at Re $\geq$ 3700, hence turbulences at the nozzle outlet. For further detailed insights, the reader is referred to e.g. [26,28]. The functional dimensions of the used nozzle were as follows:

- Inlet internal diameter is 33 mm.
- Throat internal diameter is 2 – 3.3 mm.
- Contraction ratio is ~ 10.
- Diffusion angle of the divergent section is 10°.
- Ratio of the divergent–to-convergent length is 3.

Moreover, the total flow rate inside the injector was 4.5 l/min and the static pressure at the outlet was set to 1 atm. Subsequently, the ultimate design was 3D printed and put into real operation, where the Venturi nozzle was tightly attached to the outlet of the static mixer. The deposition experiments were performed using 4 in^2 wide as-sawn Si wafers.

The simulation results demonstrated that as the gas mixture passes through the throat section, the static pressure decreases in the divergent part, according to the Bernoulli equation (4.22). See Figure 4.8. With neglecting the small temperature drop, this pressure difference causes the gas flow to deflect towards the walls, thus turbulences are generated after the nozzle outlet, which is known to occur at $Re \geq 2000$, contraction ratios higher than 7, and an angle of the divergent section lower than 22.5° [26]. In this example, the maximum flow velocity was increased from 2 mm/s at the nozzle inlet to 30 mm/s at its outlet.

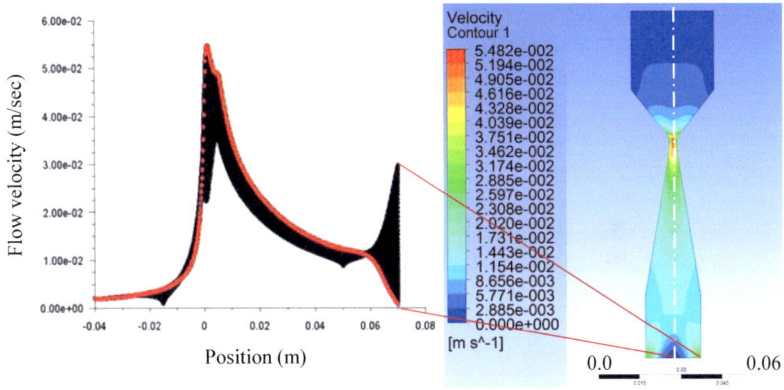

Figure 4.8: Right: ANSYS simulation chart of a cross sectional area showing the velocity contour of the gas mixture along the designed Venturi nozzle. **Left:** A graph showing the calculated velocity all over one half of the cross-sectional area. The gas stream is shown to be split into two halves increasing the velocity at the outlet compared to that at the inlet and initiating a wide distribution in the subsequent volume.

Operating the arrangement comprised of the kenics-static mixer and Venturi nozzle led to wide and uniform distribution of the gas mixture above the Si substrates. At a 15 cm-distance between the nozzle outlet and the substrate, a 1.25 in^2-wide stream at the Venturi outlet was expanded by a factor of 7 times. In comparison with the case of using only a static mixer, we eventually were able to coat a large area of about 3×3 in^2 nearly uniformly with the desired film, as demonstrated in Figure 4.9. However, this area is insufficient for coating 6-inch solar cells, thus further vortices in the injected gas mixture are required for wider distribution over the substrate surface.

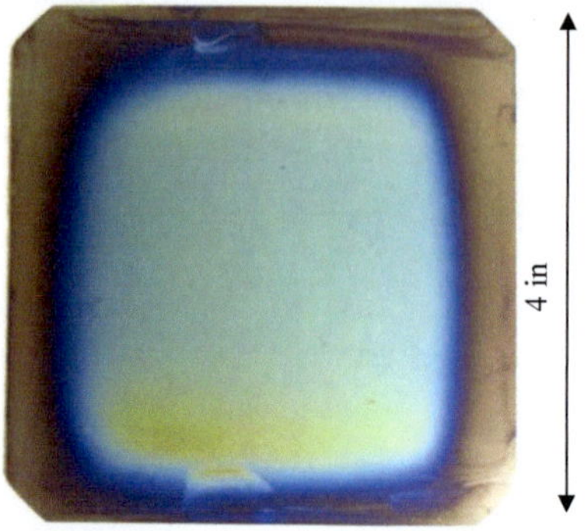

Figure 4.9: Photo taken for an APCVD SiO_{2-x} film prepared by using an arrangement of the static mixer shown in Figure 4.6 and the designed Venturi nozzle. The film deposited onto a 4 × 4 in^2 wide as-sawn Si wafer provides a coverage area of 3 × 3 in^2, nearly uniformly.

4.4.3 Numerical study of the gas distributor

This part of the injector was designed to constitute a divergent apparatus, wherein the gas mixture distributes extensively over a wide substrate area of 6 × 6 in^2 and the reactant molecules collide further. For that, a hyperbolic-shaped distributor was 3D printed after CFD analysis of the injector setup using ANSYS simulation software. The flow was modelled by the two-standard $k - \varepsilon$ equations (4.29, 30). The distributor dimensions were chosen as follows: 160 mm in height and 220 mm in diameter of the outlet opening. See Figure 4.10 (left). The inlet of the distributor was ultimately connected to the outlet of the Venturi nozzle featuring an identically inner diameter. An essential measure was inspected to obtain the highest performance of this distributor by means of obstructing the gas flow at the outlet during the deposition run. For that matter, the arrangement of the static mixer-Venturi nozzle-distributor was closed by the deposition chuck in such a way as to create a virtually sealed volume with maintaining a small gap of 100 μm. This arrangement is usually opened after the end of the deposition run in order to discharge the by-products through the exhaust. The performance of this arrangement was investigated by means of performing two simulation runs for comparison purposes, i) with an open distributor outlet, ii) and with partially closed distributor outlet. The flow boundary conditions, set in section 4.4.2, were the same in both the simulation runs.

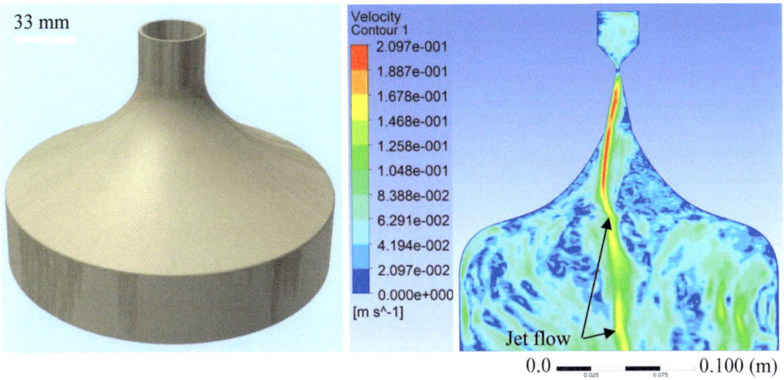

Figure 4.10: Left: A schematic picture of the hyperbolic distributor used for the APCVD injector configuration. **Right:** ANSYS simulation chart of a cross-sectional area of the Venturi nozzle-distributor arrangement featuring an open outlet to the ambient. It shows a mild flow deflection at the Venturi section and a jet flow all along the distributor without vortices.

The first simulation results depict that when there is no obstruction at the distributor outlet, a jet flow is developed all along the symmetry axis of the arrangement. Flow deflection takes place at the Venturi nozzle, yet for a short distance, after which the flow deflects back to the middle without uniform distribution all across the distributor volume. No vortices or turbulences are developed in the distributor volume indicating insufficient gas mixing above the substrates. Moreover, high-velocity flows, located closer to the outlet center, exit the distributor virtually linearly indicating a short residence time of the reactant gases before reaching the substrate surface.

On the other hand, when the distributor outlet was partially closed, deflection on both sides of the Venturi nozzle was obtained. Furthermore, vortices were formed and diffused in the distributor, by which the gas mixture expands all across the volume suggesting that the reactant molecules further collided, seen in Figure 4.11.

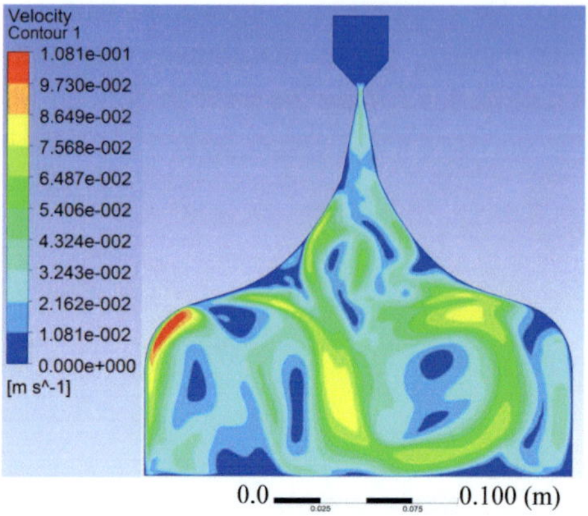

Figure 4.11: ANSYS simulation chart of a cross-sectional area of the Venturi nozzle-distributor arrangement after partially clogging its outlet showing deflections at the Venturi nozzle outlet and vortices all across the hyperbolic distributor. Decrease in the gas flow velocity takes place closer to the substrate surface promoting uniform film and high deposition rate.

The vorticity generation is well explained in [28]. In this case, the reasons behind the vortices are the turbulent characteristic of the fluid flow and the obstruction of the flow by the deposition chuck. Since the flow in the investigated volume is turbulent, the internal friction between the fluid particles is lower than their inertia. On the other hand, at the inner walls the internal friction of the fluid, also identified by fluid viscosity, is higher than the inertia. The difference between the shear stress applied on the fluid between these two regions causes dragging forces with gradual magnitude acting on the fluid from two different directions, the inner walls and the flow free region. This, in turn, leads to vortex generation in the fluid that is characterized by turbulent flow. The same effect also takes place when the fluid strikes the walls or the deposition chuck (including the mounting substrate). The fluid particles in the turbulent flow, due to the conservation of momentum, bounce back from the obstacle forming vortices. The simulation also demonstrated an increase in the flow velocity from 30 – 40 mm/s at the Venturi nozzle outlet to 80 – 100 mm/s in the middle of the distributor. The gas velocity was shown to decrease again to 60 – 70 mm/s above the substrate surface stimulating high deposition rates. This is due to the fact that the residence time of the reactants upon the substrate surface is relatively long.

Shiny etched Czochralski (CZ) Si wafers with 6-inch large area were coated with the APCVD SiO_{2-x} films using the partially closed injector arrangement. The flow rate of the $SiCl_4$ and H_2O was consistent at 2.25 l/min. Thickness mapping was performed automatically so as to assess the film thickness-uniformity by spectral ellipsometry measurement using Woollam M-2000 tool at 3 angles angle of (65°, 70° and 75°) and a wavelength range from 250 to 1000 nm. 9 observation points were chosen regularly on different positions of the coated wafer surface ensuring that the middle area and edges were included.

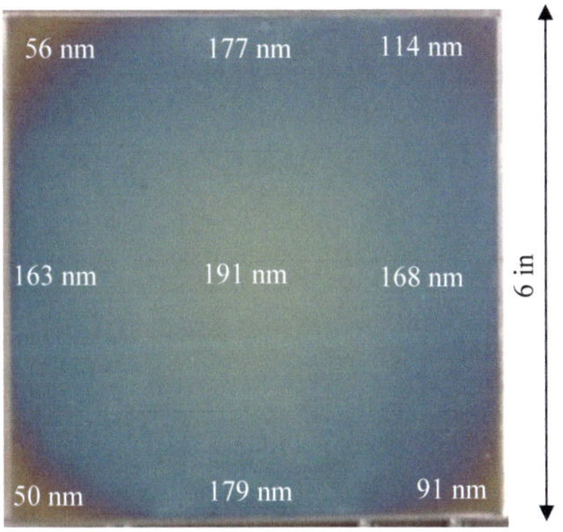

Figure 4.12: A Photo taken for a 6-inch shiny etched CZ Si wafer after coating with SiO_{2-x} film using the newly developed APCVD injector showing that 97 % of the surface area is coated nearly uniformly without thick outermost edges. The wafer corners are insufficiently coated with the oxide film.

Figure 4.12 shows a 6-inch shiny etched CZ Si wafer coated with the APCVD SiO_{2-x} film. It is exhibited that about 97 % large area of the wafer surface is coated uniformly with the SiO_{2-x} film, whereas only an area of 3 %, located at the corners, was coated insufficiently. The measured SiO_{2-x} thickness average was 176 ±14 nm, which corresponds to a variation value of ± 8 %. This covered area might be enough for coating round-edge Si wafers. However, further investigations are required to obtain a larger area of coverage with the APCVD SiO_{2-x} films and to flatten the curve of the thickness variation between the middle and the edges of the substrate. The uniform film at the wafer edges is obtained thanks to the optimization demonstrated in the next section, wherein the wafer is protected from thicker coating at the edges, comparative to the rest area.

As opposed to continuous/pulsed draining of reaction by-products, which is performed in every existing CVD/ALD technique, extremely minimized exhaustion is implemented during the deposition with the new APCVD injector. This is because of the virtually tight substrate circumference, by which only a minor volume of the out-streamed gases is permitted during deposition. This approach suggests a relatively long-term residence of the circulating gases onto the substrate surface, hence a maximum usage of the reactant portions prior to the release of by-products and a high deposition rate. The minimized gas exhaustion could be obtained by inserting 100 μm thick distance pieces between the apparatus outlet and the deposition chuck.

If the APCVD apparatus outlet is partially closed and a minor volume of the products is discharged through the exhaust, it was found that the film deposition rate is higher than that obtained when the apparatus outlet is open to the exhaust. In this case, the gas-phase diffusion in the medium was observed to decrease to a minimum extent. The evidences of that are the absolute absence of oxide particles embedded to the top surface of the deposited films, and the resulting smooth film surface. In contrast, the gas-phase diffusion increased rapidly and unintended oxide nanoparticle generation was induced when the apparatus outlet was open over the course of the deposition process. Figure 4.13 demonstrates a comparison between SiO_{2-x} films deposited onto shiny etched Float Zone Si wafers using the APCVD apparatus with the open and closed outlet arrangement.

Since hydrogen chloride (HCl) gas is produced by the $SiCl_4/H_2O$ reaction, it catalyzes the decomposition of the first intermediate products, namely the silanols, and their dehydration during the heterogeneous interaction with the substrate surface [29]. This effect takes place from the early stage of the APCVD process and is enhanced when fairly mildly gas exhaustion is permitted because of enough HCl molecules in the reactive media. No etching consequences by the HCl gas were observed on the deposited film surfaces as depicted in Figure 4.13, c. For further information, detailed investigations in accordance with the film material-quality are provided in chapter 5.

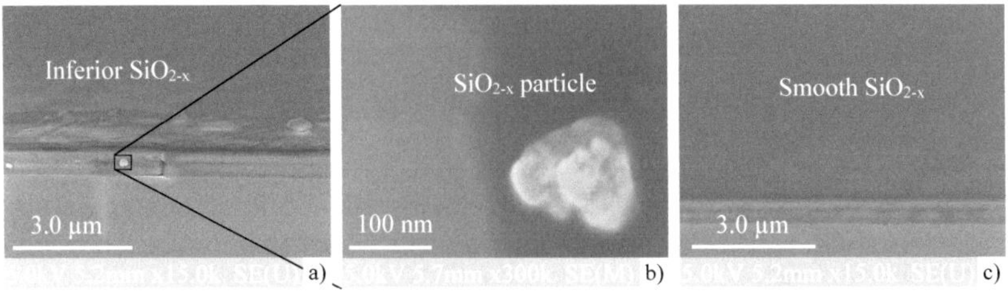

Figure 4.13: SEM images taken for APCVD SiO_{2-x} films deposited on shiny etched Float Zone Si wafers showing **a)** an inferior film structure characterized by SiO_{2-x} nanoparticles embedded to the surface after using an open-outlet apparatus, **b)** a magnification of the image a) showing a SiO_{2-x} particle, and **c)** a smooth SiO_{2-x} film adhered firmly to the Si surface without embedded particles after using a closed-outlet apparatus.

4.5 Elimination of the thick film edges

When the flowing gases inside the APCVD injector strike the substrate, irrespective of its thickness, it was observed that a narrow, yet considerably thick layer of SiO_{2-x} is prompted at the substrate edges, seen in Figure 4.14. Moreover, a wrap-around SiO_{2-x} layer is resulted at the back edges of the substrate which is unfavored when single side coating is required. These two effects are attributed to the turbulences originated at the step between the substrate and the deposition chuck and to the spikes of the gases underneath the substrate when the substrate-chuck arrangement is released.

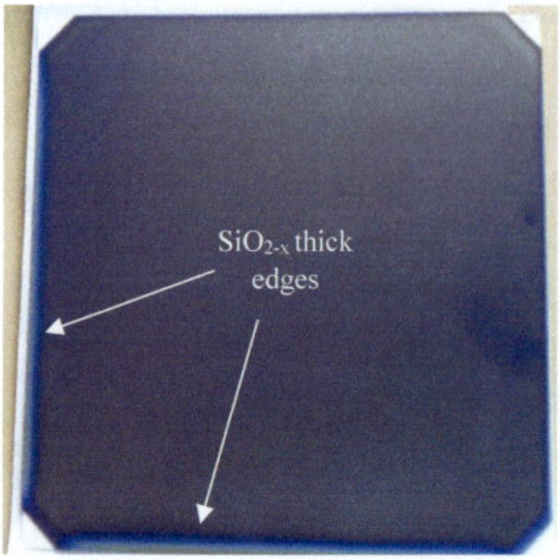

Figure 4.14: A photo taken for a 6-inch large as-sawn CZ Si wafer coated with a non-uniform APCVD SiO_{2-x} film without using a protective frame above the substrate. A narrow and thick layer is shown at the edges.

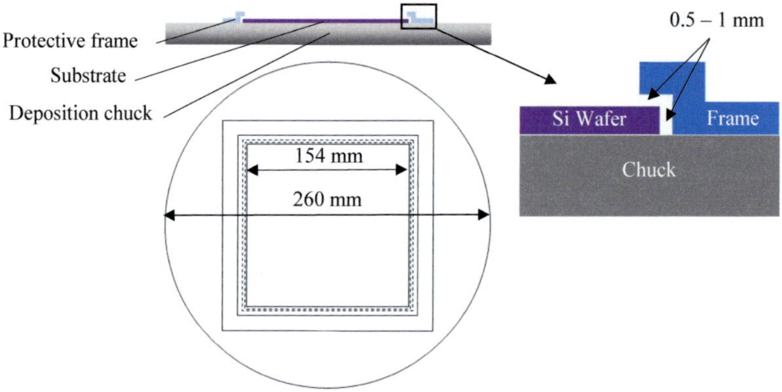

Figure 4.15: A schematic drawing of the front and horizontal projections of the deposition chuck and the protective frame on top

In order to eliminate the wrap-around layer at the substrate back edges and the thick outermost layer at the substrate front edges, the local gas turbulences and the gap between the substrate and the deposition chuck should be eliminated. For this purpose, protective frame was 3D printed and mounted above the substrates during the deposition, seen in Figure 4.15. The frame features a square shaped form and its cross-section forms a step with the substrate surface, on the internal side, and with the deposition chuck on the external side. The frame was designed in such a way as to shade less than 1 mm of the substrate edges with maintaining an upper gap of 0.5 – 1 mm.

Meanwhile, its vertical walls allow for a tolerance of 0.5 – 1 mm around the substrate so that a contact-free arrangement between the substrate and the protective frame was obtained, hence ensuring no damaged sample. It can be argumented that the frame enabled reflecting the turbulences locally at the substrate edges with maintaining large size eddies to distribute the gas mixture all over the unprotected substrate area. Consequently, coating with SiO_{2-x} films was achieved without thick rims at the outermost edges seen in Figure 4.12. To address the wrap-around influence, the deposition chuck was made of a round-shaped polycarbonate plate featuring an extremely smooth surface. Therefore, the gap between its upper surface and the substrate's bottom surface was eliminated to a further extent. This way the reactive gases could not spike underneath the substrate, hence suppressing the deposition on the substrate's back side/edges. The chuck diameter and thickness are 260 mm and 20 – 25 mm, respectively. Further details about the investigation of wrap-around elimination are available in chapter 6.

4.6 Dust formation and novel dust-proof walls

Apart from the gas-phase reaction that was proved to deteriorate within the injector volume, the surface reaction was permitted so that the APCVD SiO_{2-x} film is obtained upon the substrate surface. However, this reaction leads to undesired SiO_{2-x} particle generation upon the inner walls of the Venturi nozzle, distributor apparatus, deposition chuck, and the static mixer and its associated segments. The oxide particles were shown to fall off causing numerous defects in the films deposited upon the substrates. By way of examples, cracks, voids, porosity, low step coverage conformity, and inferior uniformity. The reader is referred to read more in chapter 5.

A novel approach was developed to prevent the SiO_{2-x} particle/dust formation and thus increase the film deposition rate upon the substrate surface. For this purpose, termination of one of the reactant molecules, such as H_2O, was on the unintended surface was realized. This was made possible by means of applying a hydrophobic coating to the unintended surface, thereby the hydrolysis of $SiCl_4$ molecules can be selectively eliminated, hence ensuring no parasitic SiO_{2-x} dust formation. Moreover, a reduction in the required reactant precursors is expected. This approach, the utilized materials, and the coating method are explained elsewhere. A proof of concept was investigated through an experimental compression between two empty tubes made of polycarbonate plastic. One tube was coated with a hydrophobic layer whereas the other tube was without coating. Both tubes were connected to two reactant sources, H_2O and $SiCl_4$ vapors, so that they intermix throughout the tube length and form SiO_{2-x} under atmospheric conditions. The operating boundary conditions were as follows: i) the average of $SiCl_4$ and H_2O evaporation rates were about 1 mg/s and 0.34 mg/s, respectively, ii) the $SiCl_4$-to-H_2O molar ratio was 0.313, iii) the relative humidity and temperature into the humidity mixer were 44.9 % and 23.1 °C, respectively, and iv) the flow rates of the $SiCl_4$ and N_2+H_2O streams were 2.18 and 1.51 l/m, respectively. To assess the effect of the hydrophobic coating, the tube weights were measured and compared before and after gas injection for 50 min in each tube using a highly sensitive scale. An endoscope camera was employed to visualize the change in the tube interiors after the injection experiments.

It was found that the weight of the tube without a hydrophobic coating increased by 370 mg after 50-min gas injection as a consequence of developing SiO_{2-x} deposits onto the inner wall. As for the tube with a hydrophobic coating, absolutely no change in the weight was measured after 50-min gas injection. Figure 4.16 (left) reveals oxide flakes developed onto the tube inner walls. This is due to the fact that the inner wall surface was continuously covered with condensed H_2O molecules, which in turn induces hydrolysis of the $SiCl_4$ flowing on top and hence forming intensive SiO_{2-x} deposits. Figure 4.16 (right), on the other hand, depicts an entire absence of SiO_{2-x} deposits on the tube inner walls. This is explained by the dehydration of the tube inner walls due to the hydrophobic coating, which suppresses the presence of OH species. Consequently, the $SiCl_4$ hydrolysis is eliminated on the wall surface and no heterogeneous reaction occurs between the intermediate products and the wall surface. This approach enabled restricting the SiO_{2-x} APCVD on the intended surfaces without the risk of parasitic deposition.

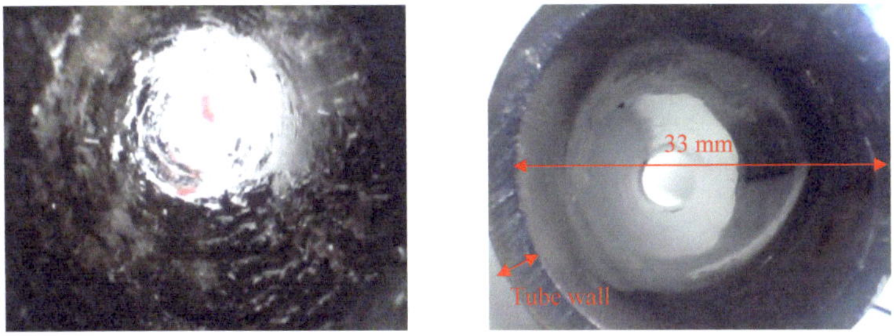

Figure 4.16: Photos taken perpendicularly to the cross section of two tubes made of polycarbonate plastic with an internal diameter of 33 mm after 50 min injection of the $SiCl_4$-H_2O mixture under atmospheric conditions, **(left)** without a hydrophobic coating onto the inner walls showing abundant SiO_{2-x} flakes and **(right)** with a hydrophobic coating showing an absence of SiO_{2-x} flakes onto the inner walls.

4.7 Advantages of the new APCVD apparatus design

1) Ability to intermix the reactant gases up to 20 cm far from the substrate without the risk of gas-phase reaction within the reactor volume.
2) Possibility to obtain a homogeneous gas mixture in front of the substrate surface with motionless blending parts.
3) Eliminated SiO_{2-x} dust formation upon the unintended surfaces such as the inner walls of the APCVD reactor.
4) Reduced periodical maintenance (cleaning) of the APCVD reactor, hence fewer operation pauses and higher operation throughout.
5) No need for a curtain flow to prevent the gas-phase reaction, as well as, for a linear substrate movement to coat the entire surface.
6) Maximized consumption of the injected reactant portions, hence high deposition rates and low wastage of freshly dosed reactants.
7) Cost-effective and ability to obtain tailored designs as the setup is made of thermoplastic materials using 3D printing technology.

8) The precise control of reactant concentrations in the gas-flow ensures constant deposition rates with no degradation in performance

9) Smooth SiO_{2-x} films without parasitic particles embedded or laying on the top surface.

10) Reasonable coverage and film thickness-uniformity exhibited all over a surface area of 6×6 in^2.

4.8 Summary

A SiO_{2-x} APCVD injector to operate at room temperature was designed and 3D printed out of thermoplastic materials. The injector was constituted of Kenics static mixer, Venturi nozzle, hyperbolic gas distributor, and deposition chuck. The Kenics static mixer which comprises 8 mixing elements was employed to intermix the two gas precursors $SiCl_4$ and H_2O sufficiently before introduction to the substrate surface. The Venturi nozzle was first investigated numerically using ANSYS CFD simulation software. A suitable design of the Venturi nozzle was found to develop deflections and backstreams in the flowing gases at the outlet. The hyperbolic distributor was first simulated by ANSYS software and the ultimate design was subsequently 3D printed. The simulation demonstrated vortex generation in the flowing gas mixture within the distributor volume. These vortices were found to perfectly homogenize and uniformly distribute the gas mixture over a 6×6 in^2 large substrate area. This was made possible because of the partial clogging of the distributor volume by the deposition chuck, by which the product exhaustion was minimized to a further extent and the residence time of the reactive gases was elongated. Because of the turbulences and obstruction by the deposition chuck and distributor walls, eddies were generated into the gas mixture circulating within the distributor. Consequently, the gas mixture was distributed over an area of less than 6×6 in^2 and the deposited films were relatively uniform with an average thickness variation of 8 %. The SiO_{2-x} films deposited with the newly developed APCVD apparatus were shown to feature a smooth surface without embedded particles. Moreover, the gas-phase transformation which is common in conventional APCVD injectors was declined to a further extent, hence reliable film quality. To protect from the effect of turbulences at the substrate edges, a frame made of 3D printed thermoplastic was installed over the substrates. This frame enabled the elimination of the thick narrow films at the substrate edges. A novel solution to avoid oxide dust formation onto the apparatus's inner walls was demonstrated, as well. A hydrophobic coating was applied onto the inner walls, by which the unintended oxide particle generation was prevented successfully.

4.9 Bibliography

[1] J. liang Zhang, J. Li, L. ma Luo, Y. hua Wo, Microstructure and morphology of SiO_x film deposited by APCVD, J. Alloys Compd. 469 (2009) 535–538. https://doi.org/10.1016/j.jallcom.2008.02.014.

[2] B. Hannebauer, F. Menzel, The combustion of $SiCl_4$ in hot O_2/H_2 flames, Zeitschrift Fur Anorg. Und Allg. Chemie. 629 (2003) 1485–1490. https://doi.org/10.1002/zaac.200300053.

[3] B.S.M. Kretzschmar, E. Wendler, A. Heft, R. Köcher, C. Voigt, C. Ronning, B. Grünler, E. Rädlein, Comprehensive porosity determination of combustion-deposited SiO_x thin films and correlation with FTIR signal, Surf. Coatings Technol. (2019). https://doi.org/10.1016/j.surfcoat.2019.07.020.

[4] H. Nagel, A. Metz, R. Hezel, Porous SiO_2 films prepared by remote plasma-enhanced chemical vapour deposition - a novel antireflection coating technology for photovoltaic modules, Sol. Energy Mater. Sol. Cells. 65 (2001) 71–77. https://doi.org/10.1016/S0927-0248(00)00079-9.

[5] J. Hee Lee, C. Hyun Jeong, J. Tae Lim, N. Gil Jo, S. Jin Kyung, G. Young Yeom, Characteristic of SiO_2 Films Deposited by Using Low-Temperature PECVD with $TEOS/N_2/O_2$, J. Korean Phys. Soc. 46 (2005) 890–894. http://spl.skku.ac.kr/_res/pnpl/etc/2005-07.pdf.

[6] A. Boogaard, A.Y. Kovalgin, I. Brunets, A.A.I. Aarnink, J. Holleman, R.A.M. Wolters, J. Schmitz, Characterization of SiO2 films deposited at low temperature by means of remote ICPECVD, Surf. Coatings Technol. 201 (2007) 8976–8980. https://doi.org/10.1016/j.surfcoat.2007.04.039.

[7] C.R. b. Kobrin, JD. Chinn, R. Nowak, Controlled vapor deposition of multilayered coatings adhered by an oxide layer, U.S. Patent 7,776,396, 2010.

[8] S.M.G. Jason W. Klaus, Ofer Sneh, Growth of SiO_2 at room temperature with the use of catalyzed sequential half-reactions, Science (80). 278 (1997) 1934–1936.

[9] J.D. Ferguson, A.W. Weimer, S.M. George, Atomic layer deposition of SiO_2 films on BN particles using sequential surface reactions, Chem. Mater. 12 (2000) 3472–3480. https://doi.org/10.1021/cm000313t.

[10] A.M. Mahajan, L.S. Patil, J.P. Bange, D.K. Gautam, Growth of SiO_2 films by TEOS-PECVD system for microelectronics applications, Surf. Coatings Technol. 183 (2004) 295–300.

[11] G. Smolinsky, R.E. Dean, LPCVD of silicon oxide films in the temperature range 410 to 600°C from diacetoxyditertiarybutoxysilane, Mater. Lett. 4 (1986) 256–260.

[12] N. Chawla, R. Nagarajan, E. Bhattacharya, Experimental and Theoretical Investigation of Thermodynamic and Transport Phenomena in Polysilicon and Silicon Nitride CVD, ECS Trans. 19 (2019) 53–68. https://doi.org/10.1149/1.3248349.

[13] A. Sherman, Plasma-assisted chemical vapor deposition processes and their semiconductor applications, Thin Solid Films. 113 (1984) 135–149. https://doi.org/10.1016/0040-6090(84)90022-1.

[14] M. Izu, T. Ellison, Roll-to-roll manufacturing of amorphous silicon alloy solar cells with in situ cell performance diagnostics, Sol. Energy Mater. Sol. Cells. 78 (2003) 613–626. https://doi.org/10.1016/S0927-0248(02)00454-3.

[15] A. Illiberi, B. Kniknie, J. Van Deelen, H.L.A.H. Steijvers, D. Habets, P.J.P.M. Simons, A.C. Janssen, E.H.A. Beckers, Industrial high-rate (~14 nm/s) deposition of low resistive and transparent ZnO_x:Al films on glass, Sol. Energy Mater. Sol. Cells. 95 (2011) 1955–1959.

[16] D.M. Dobkin, S. Mokhtari, M. Schmidt, A. Pant, L. Robinson, A. Sherman, Mechanisms of Deposition of SiO_2 from TEOS and Related Organosilicon Compounds and Ozone, J. Electrochem. Soc. 142 (1995) 2332–2340. https://doi.org/10.1149/1.2044296.

[17] https://schmid-group.com/en/business-units/thermal-processing/apcvd-system/, (n.d.).

[18] Y. Ikeda, Chemical vapor deposition method for forming SiO_2, U.S. Patent 5,462,899, 1995.

[19] Robert D. Zucker and Oscar Biblarz, Fundamentals of gas dynamics, John Wiley & Sons, New Jersey, 2002.

[20] Pijush K. Kundu and Ira M. Cohen, Fluid Mechanics, Academic Press, San Diego, 2002.

[21] R.W. Johnson, Handbook of fluid dynamics, CRC Press, London, 2016.

[22] S. Nazarenko, Fluid dynamics via examples and solutions, CRC Press, London, 2014.

[23] C. Pozrikidis, Fluid dynamics: theory, computation, and numerical simulation, Springer, New York, 2016.

[24] D.M. Hobbs, P.D. Swanson, F.J. Muzzio, Numerical characterization of low Reynolds number flow in the Kenics static mixer, Chem. Eng. Sci. 53 (1998) 1565–1584. https://doi.org/10.1016/S0009-2509(97)00132-2.

[25] D.M. Hobbs, F.J. Muzzio, Optimization of a static mixer using dynamical systems techniques, Chem. Eng. Sci. 53 (1998) 3199–3213. https://doi.org/10.1016/S0009-2509(98)00115-8.

[26] J.X. Zhang, Analysis on the effect of venturi tube structural parameters on fluid flow, AIP Adv. 7 (2017). https://doi.org/10.1063/1.4991441.

[27] HK Versteeg and W. Malalasekera, An introduction to computational fluid dynamics: the finite volume method, Pearson education, London, 2007.

[28] J.E. Meisner, R.F. Rushmer, Eddy Formation and Turbulence in Flowing Liquids, Circ. Res. 12 (1963) 455–463. https://doi.org/10.1161/01.res.12.5.455.

[29] J.G. and R. Warncke, Zur Hydrolyse von Halogeniden. I. Die Hydrolyse des Siliciumtetrachlorids, Zeitschrift Für Anorg. Chemie. 259 (1949) 109–120.

5 Analytical study of the APCVD SiO_{2-x} film quality

5.1 Introduction

Functional chemical vapor deposited (CVD) SiO_x/SiO_2 films have found numerous applications in different disciplines worldwide. By way of examples, as waveguides in integrated optics [1], as interlayer dielectrics and gate oxides in microelectronic industry such as transistor and integrated circuit manufacturing [2,3], as anti-reflection coatings, plated metallization masks, and dopant source for Si (when adding B or P) in solar cell fabrication [4–6], etc. This is mainly due to the ability of fine tuning the physical and chemical properties from inferior to virtually identical to the thermally formed SiO_2 films e.g. by Si oxidation. For instance, wide bandgap (BG = 8.9 eV), high dielectric strength (DS = 10^7 V/cm), high thermal stability and low thermal conductivity (TC = 3.2×10^{-3} W/cm·K), variable density (ρ = 2.09 – 2.49 g/cm^3), and refractive index (n = 1.11 – 1.48) suitable for multiple waveguiding structures. The chemical precursors often exploited to deposit such characteristic films are hydrosilane compounds and organics like trimethylchlorosilane (TEOS). The most common CVD techniques are plasma enhanced CVD (PECVD), low pressure CVD (LPCVD), atmospheric pressure CVD (APCVD), and combustion CVD (C-CVD), as well as, atomic layer deposition (ALD). PECVD stands out from other techniques when substrate temperatures required are relatively low (down to room temperature) and the deposition rates are high, whereas so does LPCVD when conformal step coverage is essential e.g. for microelectronic devices. SiO_x APCVD, on the other hand, is favorable when inexpensive SiO_x films are desired with similar quality and deposition rate to the PECVD films. However, for quality restrictions, the utilization of pure APCVD SiO_x films is still limited when it comes to solar cell fabrication.

Despite the fact that there has been limited information about the resilience of thus-far-available APCVD SiO_x films for photovoltaic (PV) applications, inferior quality has been often reported for other applications as follows:

1- Incorporated impurities such as hydrogen bonded groups and H_2O molecules in the films [7].

2- SiO_x particles adhered upon the film surface inducing rough surface morphology and poor step coverage conformity [8,9].

3- High void concentration and pin-holes are developed, hence low mass density, low refractive index, high etch rate, and poor electrical properties [7–10].

4- In spite of the fact that they are desired for particular applications, non-stoichiometric films are yielded when low deposition temperatures (down to 400 °C) and low reactant flow rates are setup [8–10].

5- Residual stress (200 MPa) is induced at deposition temperatures lower than 300°C in the $TEOS/O_2/O_3$ regime, while thermal stress is developed at deposition temperature of 700 °C in the SiH_4/O_2 regime [11–14].

Post-treatment of the CVD SiO_x films is often executed for remediation or functionalization purposes. This is made possible by processes such as annealed densification, gettering, reflow, planarization, and surface modification (e.g. by Si-H dangling bond introduction, graphite or boron/phosphor doping). For that high temperatures between 700 and 1200 °C are used adding up to the overall thermal budget of the APCVD process [7].

As can be noted, the properties and quality of CVD/APCVD SiO_x films strongly depend on the deposition boundary conditions and reactant materials. Since it was initiated for the first time in this research, no information has been provided about properties of the SiO_{2-x} films synthesized by the hydrolysis of $SiCl_4$ under atmospheric conditions. It is, therefore, necessary to investigate the deposition characteristics so that functional oxide films can be produced to serve various PV applications, even wider applications e.g. optics and microelectronics. The aim of this investigation is to obtain variable film properties with the newly-developed SiO_{2-x} APCVD system, hence multiple applications. More importantly, to obtain an SiO_{2-x} film quality aligning with that of thermally prepared ones e.g. prepared with other deposition technologies. For that, a fine tuning of the deposition boundary conditions is to be performed, which is quite possible for this new APCVD system, as it was shown in chapter 3. Films with high deposition rates and moderately high etch rates are, for instance, favorable when the precursor substrate is endurable to high annealing temperatures, while the film density can be tolerable and treatable (note that densities are reversely proportional to deposition rates). This is suitable for etchant mask applications, wherein the film should enable reliable protection and, on the other hand, fast etching process. In contrast, films formed at low deposition rates, regardless of their thickness, are favorable when processed samples are sensitive to high annealing temperatures. For that the film density should be efficiently high. This is suitable for electroplating mask (e.g. in SHJ solar cells) and intermetal dielectric applications.

In this chapter an analytical study is carried out to trace the impacts on the SiO_{2-x} film properties, namely the deposition parameters (gas flow rate, reactant concentration and molar ratio, and substrate temperature), as well as, the post-deposition treatment. Furthermore, the significance of adjusting the deposition parameters is exhibited for both film quality and functions. First, the nature of the deposited oxide film was inspected so as to assure the interaction yield between H_2O and $SiCl_4$ reactants. Second, the influence of the deposition parameters on the film deposition rate was investigated. Third, structural defects were traced at various film thickness and substrate surface topographies, by which compatibility with PV application. Forth, effects of variable deposition parameters on the chemical properties of the oxide film such as purity, atomic stoichiometry and etch rate. Fifth, influence of variable deposition parameters on the film optical constants such as index of refraction, extinction attenuation coefficient, surface roughness, and reflectance/ transmittance. Sixth, investigating annealing treatment impacts on film structure and durability, thickness, chemical properties, and optical properties. The wide range of deposition conditions and provides variety of oxide film characteristics. The calcination experiments demonstrate a thorough effect on the film integrity and stoichiometry, by which it can withstand harsh environments such as acidic solutions.

5.2 Inspection of the film-nature

As an essential inspection, identification of the chemical nature of the $SiCl_4/H_2O$ films was carried out. For that, two sets of 2×2 cm^2 wide and 200 µm thick substrates were used; shiny etched Float Zone (FZ) and smooth Ge wafers. The FZ Si wafers were cleaned before deposition using the so-called Radio Corporation of America (RCA) [15]. The whole wafers were coated with about 150 nm thick APCVD SiO_{2-x} films at room temperature without a subsequent treatment. The deposition conditions are summarized in Table 5.1.

Table 5.1: The APCVD deposition parameters used to inspect the chemical nature of the silicon oxide films

Input parameter	Value
N_2 flow rate in the H_2O bubbler	2.2 l/min
N_2 flow rate in the humidity mixer	0.4 l/min
N_2 flow rate in the $SiCl_4$ vaporizer	0.1 l/min
Absolute humidity at the injector inlet	16 g/m^3
$SiCl_4$ concentration at the injector inlet	200 g/m^3
APCVD system temperature	25. 2 °C

First, the oxide-coated FZ Si samples were subjected to chemical composition analysis using Fourier transform infrared spectroscopy (FTIR) inspection. This experiment was performed to indicate the chemical bonds and possible impurities incorporated into the film volume. The tool used was VERTEX 80v FTIR Spectrometer from Bruker supplied with OPUS software. The measurements were performed under vacuum at room temperature without N_2 cooling. The spectral range was from 400 to 4400 cm^{-1}. The resolution was set to be between 4 and 8 cm^{-1}, at which the relative amount of noise in the spectra was reduced to a further extent. This is due to the enough intensity of

IR light passing through the aperture and entering the detector. The number of iterations was set to 10.

Second, for further proof, the oxide-coated Ge samples were subjected to elemental analyses by means of energy dispersive X-ray spectroscopy (EDX) inspection. This experiment was intended to analyze the entire volume of the APCVD SiO_{2-x} film. For that a substrate material other than Si was to be employed. Unlike Si wafers, choosing Ge as a substrate material was beneficial for avoiding interference by Si atoms in the EDX spectra. Moreover, the film thickness of 150 nm was smaller than the penetration depth of the EDX electron beam. The used instrument was Bruker Quantax 400 in a Zeiss Auriga 60 scanning electron microscope (SEM). The samples were positioned in the SEM chamber as though the electron beam strikes their surface perpendicularly, the acceleration voltage was adjusted at 5 kV, the working distance to the surface was 5.5 mm, and the penetration depth was 300 nm.

A chemical bonding configuration of an APCVD oxide film was obtained by the FTIR measurements. As shown in Figure 5.1, various absorption bands attributed to chemicals composed of Si, O and H are present as follows [16–18]:

i) Si-O-Si rocking band is positioned at the wavenumber 447 cm^{-1}, Si-O-Si bending band is positioned at 798 cm^{-1}, and two Si-O-Si stretching bonds are positioned at 1072 and 1160 cm^{-1}.
ii) Si-O-H bonds are positioned at 917, 3370 and 3664 cm^{-1}. Usually these bonds are attributed to impurities in the silicon oxide films such as silanol molecules and hydroxyl (OH) groups.
iii) No obvious H2O molecules nor Si-H groups appear in the FTIR spectra as shown at the wavenumbers 1678 and 2300 cm-1. The presence of OH groups is an initial indication of slightly non-stoichiometric characteristics of the silicon oxide films. The elimination of the parasitic OH groups and the modification of the film stoichiometry are investigated later in this chapter. Meanwhile, Cl bonds are not found in this representative spectrum. This suggests the absence of Cl groups in the films or they are below the IR absorption limits.

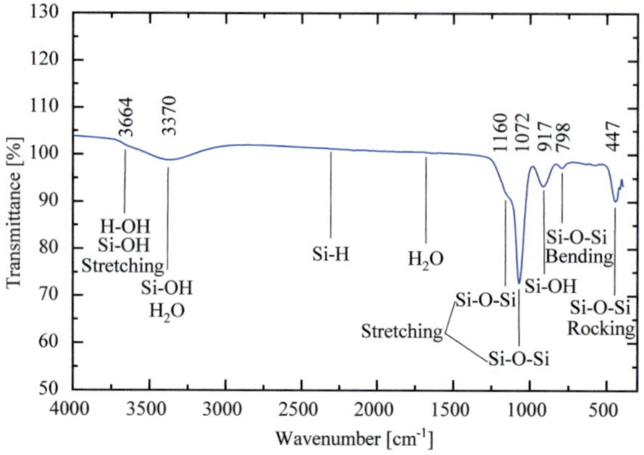

Figure 5.1: FTIR spectra of an inspected APCVD SiO_{2-x} coated FZ Si wafer illustrating functional groups corresponding to the silicon oxide network at well-defined IR wavenumbers

The EDX investigation of coated monocrystalline Ge samples resulted in the spectra shown in Figure 5.2. Three main elements were detected: Si, O and Ge, in addition to a negligible amount of C (expected from possible contamination when handling the samples). The presence of Ge element is an indication of inspecting the entire silicon oxide film cross-section. A calculation of the element contributions was automatically performed by the software provided by Bruker for quantification of measured EDX spectra. The mass calculation revealed, after normalization, 43.3 at % of Si, 46.5 at % of O and 9.8 at % of Ge. These values demonstrate that despite the high penetration depth of the EDX electron beam, the contribution of Ge was relatively small, whereas most of the detected volumes were Si and O. The EDX analyses provided a rough estimation of the stoichiometric degree of the oxide films, as well. The Si:O atomic ratio calculated from the spectra (based on the atomic weight of the elements) was 1:1.8. This indicates that the as-deposited oxide films are sub-stoichiometric and therefore we explicitly used the formula SiO_{2-x} to abbreviate the APCVD films throughout this research. However, the measuring errors of the O and Si atomic concentrations was as high as ± 20 %. This measurement error was, nevertheless, unaffected by the small amount of Ge. Annealing of the deposited film is proven, later in this chapter, to improve its stoichiometry to the degree aligning with thermally prepared SiO_2 films. For result confirmation, X-ray photoelectron spectroscopy (XPS) investigation for the same samples is available in the appendix.

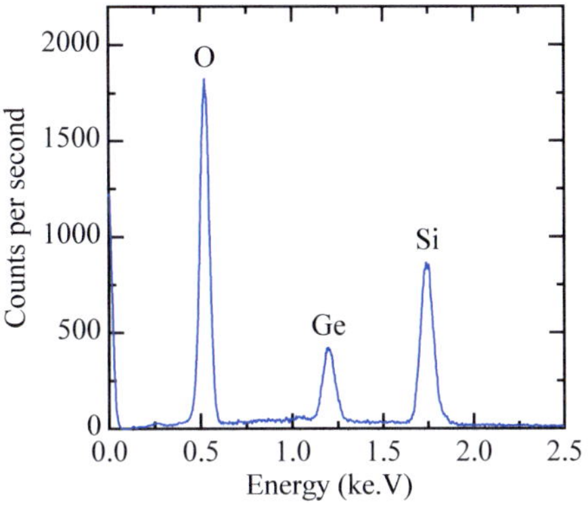

Figure 5.2: EDX spectra demonstrating the element analyses of as-deposited 150 nm thick SiO_{2-x} films prepared with the APCVD system at room temperature on smooth Ge samples

5.3 Impact of deposition boundary conditions on the SiO_{2-x} deposition rate

5.3.1 Flow rate pattern and reactant concentration

The flow rate of the gas precursors was proven to extensively influence their evaporation rate and concentration into the carrier gas volume, as represented in chapter 3. According to prior arts, this suggests to have impacts on the deposition rate of the APCVD SiO_{2-x} films [10,12]. In this section the correlation between flow rates of both $SiCl_4$ and H_2O vapors and SiO_{2-x} deposition rate was investigated so that a careful design of the flow patterns of the SiO_{2-x} APCVD process can be obtained. Furthermore, the relationship between reactant concentrations into the active moiety mixture introduced to the APCVD injector and the film deposition rate was assessed. Therefore, the following experiments were performed.

First, the effect of the H_2O vapor parameters on the SiO_{2-x} deposition rate was investigated. For that the flow rate of the carrier gas N_2 through the water bubbler (Q_b) was varied between 0.24 to 1.41 l/min at a constant temperature of $T_b = 26$ °C. The flow rates of N_2 into both the humidity mixer (Q_m) and the $SiCl_4$ evaporator (Q_{SiCl4}) were maintained at 1.3 l/min. The temperatures into the humidity mixer (T_m) and $SiCl_4$ evaporator (T_c) were measured at 27 °C. A set of shiny etched FZ Si wafers with an area of 2×2 cm^2 were employed. They were cleaned by the standard HNF[1] method. Subsequently, the wafers were coated with APCVD SiO_{2-x} films at the gradual Q_b values. All deposition runs were fixed at 5-min duration. Eventually, the thickness of the deposited films was measured and the deposition rate (DR), evaluated per min, was calculated. For thickness measurements Woollam M-2000 spectral ellipsometer was used at one angle of 70° and a wavelength range from 250 to 2000 nm.

Figure 5.3 (left) shows that the deposition rate of the APCVD SiO_{2-x} film increases linearly with the increase of Q^1_b. However, these results were not found enough for adjusting the deposition rate of the films as long as this rate changes at other bubbler temperatures at the same tested Q_b values. This is due to the fact that the absolute humidity in the mixer (AH_m) changes at various temperatures in the water bubbler (more details were shown in chapter 3). Since the variation of T_b is essential in the newly developed SiO_{2-x} APCVD system, the effect of the AH_m change on the deposition rate was to be investigated.

At the same temperatures mentioned in the previous set of experiments, a variation of the SiO_{2-x} DR at variable AH_m from 2 to 22 g/m^3 was performed. For that the Q_m and Q_b were calibrated for every AH_m value. For reproducibility, the deposition experiments were carried out at three different concentrations of the $SiCl_4$ vapor ($Kc_{(SiCl4)}$) into the relevant evaporator: 80, 120 and 180 g/m^3. The duration of all deposition runs was fixed at 6 min. The thickness of the deposited films was measured using ellipsometry measurements and the SiO_{2-x} DRs were subsequently calculated.

[1] HNF is a cleaning method consists of the first stage of the RCA cleaning – the HNO_3/H_2O cleaning, flowed by an oxide strip using HF/H_2O solution.

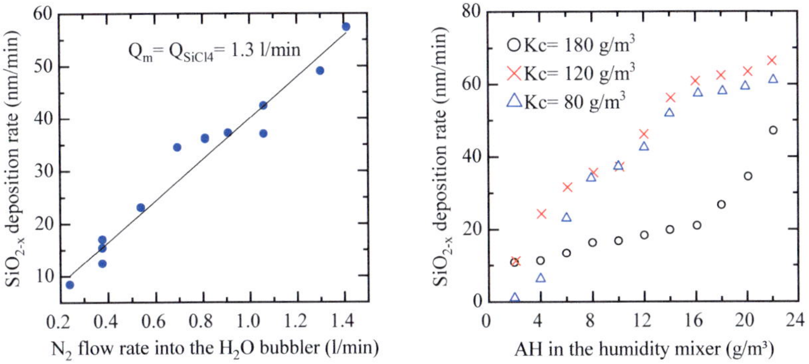

Figure 5.3: Graphs showing the APCVD SiO_{2-x} deposition rate as a function of (**left**) the gas flow rate introduced to the water bubbler at $Q_m = Q_{SiCl4} = 1.3$ l/min and $T_b = T_m = T_c = 26$ °C, and of (**right**) the water vapor concentration into the humidity mixer at three gradual $Kc_{(SiCl4)}$ and $T_b = T_m = T_c = 26$ °C

As depicted in Figure 5.3 (right), the SiO_{2-x} DR increases with the increase of AH_m. It is to be noted that at low $Kc_{(SiCl4)}$ of 80 to 120 g/m³, the DR curves are almost flattened after the AH_m of 16 g/m³ and the increase in DR diminishes. Vice versa, the curve at $Kc_{(SiCl4)}$ of 180 g/m³ is in continuous increase even after AH_m 22 g/m³. The interpretation of these two different curve behaviors is that at low $Kc_{(SiCl4)}$ within the working vapor mixture its whole volume is consumed and the H_2O vapor is excessive. On the other hand, at large $Kc_{(SiCl4)}$ the increasing AH_m higher than 16 g/m³ can still intermix with enough $SiCl_4$ volume, wherein the mixture is in excess of $SiCl_4$ vapor. Consequently, the impact of AH_m overshadows the Q_b impact as it allows for reliable fine tuning of the DR irrespective of the Q_b and T_b. It is worth pointing out that the curve related to $Kc_{(SiCl4)} = 180$ g/m³ is lower than those of small $Kc_{(SiCl4)}$ values. The reason for lower DRs at high is $Kc_{(SiCl4)}$ elaborated in the next experiments.

Second, the effects of the $SiCl_4$ vapor boundary-conditions on the SiO_{2-x} deposition rate was systematically investigated. A set of 2×2 cm² wide shiny etched FZ Si wafers were employed after cleaning by the standard HNF method. Q_b and Q_m were stabilized at 0.38 and 1.71 l/min, respectively. This provided an absolute humidity at the injector inlet of $AH_m = 4$ g/m³. The whole deposition system-temperature was stabilized at $T_{APCVD} = 23$ °C. Q_{SiCl4} was tuned between 0.02 and 2.43 l/min with a logarithmic interval. At every Q_{SiCl4} an Si sample was coated with APCVD SiO_{2-x} film for 14 min each, and the $Kc_{(SiCl4)}$ was measured each deposition run. Ultimately, the thickness of the deposited films was measured by ellipsometry inspection, using Woollam M-2000 spectral ellipsometer at one angle of 70° and a wavelength range from 250 to 2000 nm.

In agreement with the finding in chapter 3, as the Q_{SiCl4} increases, the $SiCl_4$ evaporation rate (E_{SiCl4}) increases and the $Kc_{(SiCl4)}$ decreases. A sharp $Kc_{(SiCl4)}$ decrease takes place at Q_{SiCl4} less than 0.21 l/min, seen in Figure 5.4 (left). Also shown in Figure 3.4 (left) is the SiO_{2-x} DR correlation with both Q_{SiCl4} and $Kc_{(SiCl4)}$. One expects that the relation between the film DR and the $Kc_{(SiCl4)}$ is linear. However, it was shown that SiO_{2-x} DR anomalously increases from 4.3 to 25 nm/min at $Kc_{(SiCl4)}$ range from 1254.5 to 97.6 g/m^3, which corresponds to the Q_{SiCl4} range from 0.02 to 1.06 l/min. Thereafter, it was found that as the Q_{SiCl4} increases the SiO_{2-x} DR decreases gradually with the continuous decline of $Kc_{(SiCl4)}$.

For better understanding, experiments to trace the reproducibility of this finding were carried out at various AH_m values: 5, 10 and 15 g/m^3. However, the flow rate of humid N_2 at the mixer outlet $Q_{(b+m)}$ (or the injector inlet) was stabilized at 2.2 l/min. The T_{APCVD} system was measured at 23 °C. At every AH_m set value the Q_{SiCl4} was varied from 0.21 l/min up to 2.01 l/min with a logarithmic interval, for each the corresponding $Kc_{(SiCl4)}$ was measured. After RCA cleaning, 2×2 cm^2 wide shiny etched FZ Si wafers undertaken an SiO_{2-x} film coating, each for 5 min. Subsequently, the deposited film thickness was measured for all samples with Woollam M-2000 spectral ellipsometer at three different angles: 65°, 70° and 75°, each at a wavelength range from 250 to 2000 nm. Henceforth, the relevant deposition rates were calculated and plotted as a function of the inverse of $Kc_{(SiCl4)}$ as revealed in the right graph of Figure 5.4.

The findings in this set of experiments confirmed with those demonstrated in Figure 5.4 (left). It was shown that as the Q_{SiCl4} increases, the $Kc_{(SiCl4)}$ decreases and the SiO_{2-x} DR increases to maximum values corresponding to Q_{SiCl4} of 0.78 – 1.41 l/min, after which the SiO_{2-x} DR diminishes to a low value corresponding to the highest Q_{SiCl4} set: 2.21 l/min. The expected reason is that at Q_{SiCl4} lower than 1 l/min the static mixer portion of the APCVD injector does not function sufficiently with its currently designed dimensions.

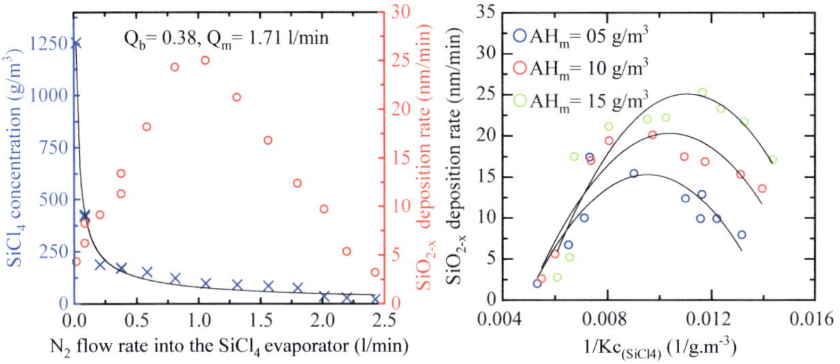

Figure 5. 4: Left: A graph showing both the $Kc_{(SiCl4)}$ and the DR of the APCVD SiO_{2-x} as a function of the gas flow rate introduced to the $SiCl_4$ evaporator (the continuous line is a guide for the eye). **Right:** A graph showing the DR of the APCVD SiO_{2-x} as a function of the inverse of $Kc_{(SiCl4)}$ at three gradual AH_m values (the fitting used is polynomial of the second degree).

The diameter of this static mixer is fairly wide for flow rates lower than 3.5 l/min for both gas streams, and thus the intermixing of the two gases is incomplete. Moreover, the low flow rate of at least one gas stream induces non-uniform distribution of the gas mixture over a wide surface area, usually designed to suit 6-inch substrates. Therefore, lower film thickness is resulted e.g. in the middle of the deposition chuck where the samples were practically positioned. In Figure 5.4 (right) it was also shown that the fitted three curves shift toward lower $Kc_{(SiCl4)}$ values (or higher Q_{SiCl4} values) as the AH_m increases. The maximum film deposition rates: 17.1, 20.1 and 25.3 nm/min were obtained at $SiCl_4$ concentrations: 136.7, 102.8 and 85.8 g/m^3, corresponding to AH_m of 5, 10 and 15 g/m^3, respectively. It was consequently deduced that considering the AH_m and $Kc_{(SiCl4)}$ individually is inefficient for the determination of SiO_{2-x} DRs and hence film properties. This therefore gave rise to adopting the molar ratio (MR) between $SiCl_4$ H_2O reactants as an accurate parameter for investigating SiO_{2-x} films.

5.3.2 $SiCl_4$-to-H_2O molar ratio

The influence of $SiCl_4$-to-H_2O MR within the working gas mixture on the SiO_{2-x} DR was studied. For that, two sets of MR variation experiments were performed as follows: the SiO_{2-x} DRs were varied at increasing $Kc_{(SiCl4)}$ values while maintaining the AH_m at 5 and 15 g/m^3, respectively. The deposition temperature was T_{APCVD} = 25 °C over the course of all the experiments. Ellipsometry measurements were carried out to gauge the film thicknesses using Woollam M-2000 spectral ellipsometer at three different angles: 65°, 70° and 75°.

Figure 5.5 depicts two different plots representing APCVD SiO_{2-x} DRs as a function of $SiCl_4$-to-H_2O MR. It is seen that as the MR increases from zero, the film DR increases gradually to a maximum value at MRs between 0.57 and 1.08. Thereafter, the DR overwhelms to a low value at MRs between 2.6 and 5.85, after which no change in the DR occurs irrespective of the further increase of MR. The low DRs obtained at MRs $\leq$ 0.57 are attributed to the low $Kc_{(SiCl4)}$ yielded at high Q_{SiCl4} values. On the other hand, the reduction of DRs at MRs $\geq$ 1.08 are directly attributed to the extremely low Q_{SiCl4} values, at which the pressure drop of the gas flow is inadequate to intermix the two vapors, although $Kc_{(SiCl4)}$ is large, as shown above. Furthermore, when the constant H_2O volume introduced to the vapor media is completely consumed, the change in DR is extremely small, irrespective of further increase of the $SiCl_4$ volume into the vapor media. As a matter of fact, the greatest APCVD SiO_{2-x} DR was shown to be obtained at $SiCl_4$-to-H_2O MRs between 0.5 and 1, considering that the stoichiometric MR between these two reactants is 0.5 in ideal cases. See equation (5.1) in the next section.

To summarize, the APCVD SiO_{2-x} DR limiting factor is the $SiCl_4$ at MRs $\leq$ 0.5, whereas it is the H_2O concentration at MRs $\geq$ 1. Additional reasoning of this finding is explained by quantum chemistry in the following section 5.3.3.

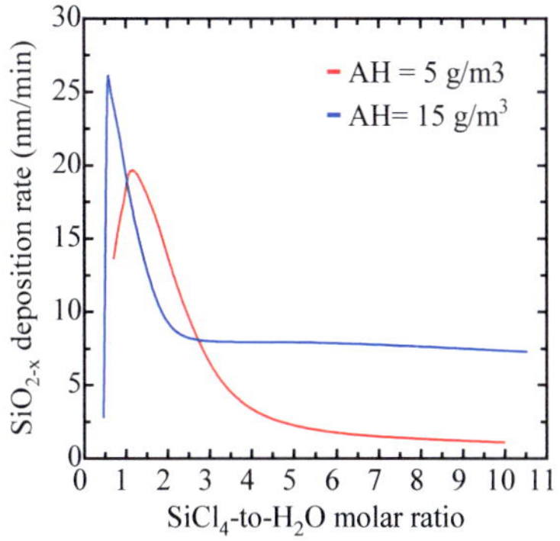

Figure 5.5: A graph demonstrating the APCVD SiO_{2-x} growth rate as a function of $SiCl_4$-to-H_2O molar ratio into the working gas mixture introduced to the injector at different AH_m and $Kc_{(SiCl4)}$ values

5.3.3 Substrate temperature

The substrate temperature (T_s), or the depositon chuck-temperature (T_{chuck}), is an extreme deposition parameter when it comes to the reaction between $SiCl_4$ and H_2O vapors. As this reaction is exothermic and takes place at or below room temperature without considerable activation energy, a small change in the T_s should leave an impact on the SiO_{2-x} deposition rates. Therefore, fine-tuning of the substrate temperature was essentially executed to define the APCVD SiO_{2-x} DR. Accordingly, the correlation between T_s and SiO_{2-x} DR was investigated. The T_s fine tuning was shown to possess more impacts on the SiO_{2-x} film characteristics, as well. This will be shown later in this chapter.

For that matter, numerous substrates made of chemically etched 2×2 wide FZ Si wafers were exploited after RCA cleaning. The substrates were divided to three groups, each of which was coated with APCVD SiO_{2-x} films at a distinctive MR. The substrates of each group were coated separately at T_s range from 278 to 323 K with an interval of 5 K. For the case of deposition runs conducted at temperatures higher than the room one, a hotplate was used to determine the T_{chuck}, whereas a deposition chuck cooled with a fluidized N_2 was used for deposition runs conducted at lower temperatures. The coated samples were characterized by ellipsometry measurements so as to figure out the film thicknesses. The heating/cooling technique is not explained in this book. Woollam M-2000 spectral ellipsometer was used at three different angles: 65°, 70° and 75° and a wavelength range of 300 - 1000 nm.

Figure 5.6 reveals three curves representing the change in the APCVD SiO_{2-x} growth rate as a function of the inverse of T_{chuck} (or T_s) at $SiCl_4$-to-H_2O MRs: 0.72, 3.2 and 4.50. It was exhibited that the DR increases with the decrease in T_{chuck}. Moreover, at $T_{chuck} \geq 333$ K no deposition occurs regardless of the MR set. On the contrary, the SiO_{2-x} DRs at $T_{chuck} \leq 290$ K are quite high and increase gradually as the T_{chuck} approaches 273 K. The curve corresponding to the MR of 0.72 is higher than the other two, referring to the highest SiO_{2-x} DRs. This is consistent with the previous conclusion indicating that at the highest SiO_{2-x} DRs are obtained at MRs between 0.5 and 1, above which the greater the MR, the lower the SiO_{2-x} DR.

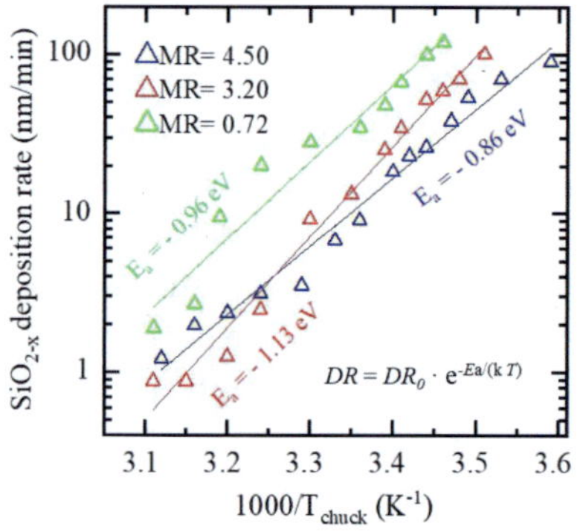

Figure 5.6: A logarithm of the SiO_{2-x} DR as a function of the inverse T_{chuck} at three different $SiCl_4$-to-H_2O MRs showing a negative activation energy of around 0.86 – 1.13 eV. The interpolation fitting is, therefore, linear.

The interpretation of this surprising result is way much more complicated than appeared, explicitly. However, in the followings, we tried to briefly simplify this phenomenon without diving into too much chemical complications, since this book is a more solar cell and mechanical engineering directed topic than chemistry. It is to consider that our focusing area is the hetero-functional condensation reaction of the reactant species with the Si surface. The gas-phase reaction is unremarkable in this research as shown in Chapter 4. When the $SiCl_4$ hydrolysis reaction is complete, its theoretically overall enthalpy-change equals (- 865 kJ/mol).

This is calculated according to the bond energies summarized in Table 5.2 [19], making use of the equations:

$$SiCl_4 + 2\ H_2O \rightarrow SiO_2 + 4\ HCl + \text{thermal energy} \quad (5.1)$$

$$\Delta H = \Delta H_{(SiCl4)} + 2\ \Delta H_{(H2O)} - \Delta H_{(SiO2)} - 4\ \Delta H_{(HCl)} \quad (5.2)$$

where:

ΔH is the total enthalpy change of the reaction (kJ/mol).

ΔH_{SiCl4} and ΔH_{H2O} are the enthalpy changes of the Si-Cl and O-H bond breaking.

ΔH_{SiO2} and ΔH_{HCl} enthalpy changes of the Si-O and H-Cl bond coordinating.

Usually, the bond coordinating is suggested to be negative and the bond breaking is accepted to be positive. It is to be considered that the energy required to break a chemical bond should be at least equal to that of the bond between its relevant atoms.

Table 5.2: The energy of the bonds between atoms present in equation 5.1 measured in kJ/mol [19]

Chemical bond	O-H	Si-Cl	Si-O	H-Cl
Bond energy (kJ/mol)	459	381	452	428

On the basis of these results and on what is frequently accepted, it is realized that the condensation reaction between $SiCl_4$ and H_2O molecules that leads to hetero-functional phase SiO_{2-x}, is spontaneous, yet slow, exothermic and thermodynamically unstable. That is why this reaction is known to be kinetically stable, hence kinetically controlled. On the other hand, T. Kudo *et al.* [20] estimated that it is possible to diminish the activation energy (E_a), also known as activation barriers, throughout all the elementary reactions by the coordination of further OH radicals to the reaction transition-state (the equations of the relevant reactions are stated in chapter 3, section 3.2).

Therefore, the rate of reaction (in this case is hetero-phase one) gets faster and thus it is converted to kinetically unstable. In other words, the more the H_2O molecules are incorporated into the reaction steps, the more thermal energy is released, which is a reaction product as well. This translates to a shift of the $SiCl_4$ hydrolysis reaction-equilibrium to the product side along with the increase of the reaction surrounding temperature. In our studied case, as the T_s decreases below 295 K, more H_2O molecules are adsorbed to the substrate surface and if the T_s is below the dewpoint temperature (T_{dp}), extremely fine H_2O droplets are formed on the substrates top surface. Therefore, the condition of additional H_2O molecules is realized in both H_2O condensation cases. As a consequence, the rate of the hetero-functional reaction increases gradually. On the flip side, the effect of T_s increase can be explained by the same interpretation, as well. As the T_s increases the H_2O molecules start desorbing from the substrate surface until they are, at T_s of 333 K, minimized to a further extent could not be detected with attenuated total reflection (ATR) FTIR measurements. Consequently, no deposition occurs irrespective of the presence of $SiCl_4$ species upon the substrate surface. This is in good agreement with what was proven in chapter 4 – when the terminal H_2O layer is eliminated upon the injector inner walls, as a consequence of the hydrophobic coating, no hetero-phase condensation reaction takes place. Accordingly, the deposition rate of unintended SiO_{2-x} drops to zero.

In accordance with the mentioned above, the effect of the MRs, corresponding to the three SiO_{2-x} DR curves shown in Figure 5.6 is well understood. The highest SiO_{2-x} DR curve obtained at the lowest MR = 0.72 is ascribed to the more OH coordination to the Si wafer top surface, which is, in turn, due to more H_2O molecules incorporated in the $SiCl_4$ hydrolysis reaction. The higher the MR is set, the less the H_2O molecules are incorporated in the reaction, and therefore the less the OH bonds are coordinated to the Si wafer top surface resulting in lower SiO_{2-x} DRs. It is to be concluded as well, that the MR values mentioned in this graph are correct only at $T_s \leq$ room temperature, at which no desorption of the H_2O molecules occurs. The higher the T_s raises, the larger the MR gets at the Si wafer top surface as a consequence of the H_2O desorption, which is translated to lower SiO_{2-x} DR.

It is worth pointing out that at high APCVD SiO_{2-x} DRs, the film characteristics are unnecessarily sufficient, e.g. for aligning with the integrity of thermally prepared films. This tread-off relation between deposition rate and film properties are studied in the subsequent sections.

5.4 Impact of APCVD boundary conditions on the SiO_{2-x} chemical properties

Reactant concentrations, molar ratio and deposition temperature were proven to have an impact on the chemical characteristics of the APCVD SiO_{2-x} films, besides the films deposition rates. In this section film chemical compositions, impurities, stoichiometry and etch rate were studied as a function of the deposition parameters.

5.4.1 Study of the SiO_{2-x} film composition and chemical bonds as a function of $SiCl_4$-to-H_2O MR

Chemical compositions of the APCVD SiO_{2-x} films were investigated at various $SiCl_4$-to-H_2O MRs by means of FTIR measurements. The MRs set were varied from below 0.5 to values higher than 1, by which the effect of different DRs was traced, as well. 2×2 cm^2 wide chemically polished FZ Si wafers with thickness of 240 ±5 μm were utilized after RCA clean. It is to be noted that the higher the substrate thickness, the better the FTIR spectra resolution and reliability. The substrates were coated, on one side, with APCVD SiO_{2-x} films at MRs range from 0.33 to 4.55 at room temperature for 6 min, each. For that, the total flow rate of the humid N_2 was maintained at $Q_{(b+m)}$ = 2.18 l/min while Q_{SiCl4} was calibrated between 0.35 and 1.56 l/min. Subsequently, the film thickness of the deposited films was first assessed using ellipsometry measurements. Woollam M-2000 spectral ellipsometer was employed at one angle of 70° and the relevant wavelength range was from 300 to 900 nm. The chemical composition of the deposited films was measured with FTIR using VERTEX 80v FTIR instrument from Bruker. The measurements were executed under vacuum at room temperature at a spectral range from 400 to 4400 cm^{-1}. The stepwise measurement was 8 cm^{-1} and the number of iterations was set to 10. The measurement duration was 60s for each sample. These measurements were compared with a measured sample of chemically polished FZ Si without APCVD SiO_{2-x} coating (except for the few nanometers native oxide). The peak subtraction, between the coated samples and the uncoated ones, was automatically conducted by the OPUS software associated with the FTIR instrument. Baseline correction of the spectra was carried out using both OPUS and Origin 2020[2] software, additionally to the peak picking. The FTIR spectra are exhibited in Figure 5.7 and the ultimate measurement results are summarized in Table 5.3.

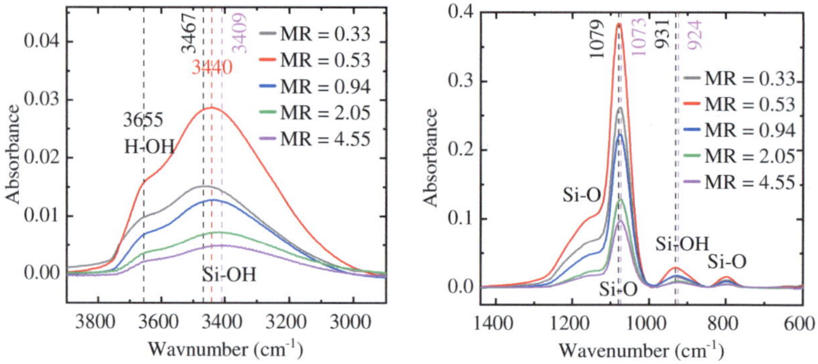

Figure 5.7: Graphs showing magnified FTIR spectra of the APCVD SiO_{2-x} films prepared at various MRs and their absorption band at (left) the wavenumber 3000 – 4000 cm^{-1} and (right) the wavenumber 600 – 1400 cm^{-1}

In Figure 5.7, the magnified spectra elucidate that Si-O, Si-OH and H-OH bonds are present in all the deposited films irrespective of the MR set [16–18]. It is shown that the intensity of the absorption bands (evaluated by peak height and broadness) are affected by the MR, as such it possesses the highest value at MR = 0.5, below and higher than which the bond density in the film decreases. This is consistent with the findings reported in section 5.3.2, since the bond density of a specific group translates to the concentration of its relevant species in the investigated film. Therefore, as the MR increases starting from zero, the film thickness increases to its highest value at MR of 0.5 and decreases again at larger MRs, see Table 5.3.

2 An Origin software is licensed for TU-Ilmenau with a Serial Number: SF8T5-6089-7609543

Table 5. 3: Bond density and thickness measurements of the deposited APVD SiO_{2-x} films at various MRs

Item	Unit	MR = 0.33	MR = 0.53	MR = 0.94	MR = 2.05	MR = 4.55
Film Thickness	nm	272.2	332.1	207	98.1	67.9
Si-OH	cm^{-1}	931	931	929	926	924
Si-OH peak height	-	0.0154	0.0286	0.0170	0.0090	0.0074
Si-O stretch	cm^{-1}	1076	1079	1076	1076	1075
Si-O peak height	-	0.2637	0.3869	0.2240	0.12876	0.0974
Si-OH stretch	cm^{-1}	3467	3440	3433	3417	3409
Si-OH peak height	-	0.0151	0.0286	0.0127	0.0072	0.0049

In terms of impurity no indications of Cl bonded groups are proven to exist. Si-Cl bonds, for instance, usually absorb at a wavenumber between 550 and 750 cm^{-1} [21,22]. Meanwhile, H-Cl groups are usually present between 2537 and 2650 cm^{-1}, since H atoms are light and their bonds are strong e.g. compared to Si [23,24]. In our case of study, the OH impurities are present at mainly three different wavenumbers: 931, 3440 and 3655 cm^{-1} [16,18]. The MR of 0.53 resulted in the most intense Si-OH absorption band positioned at a wide range of wavenumbers (almost 2950 - 3850 cm^{-1}), see Figure 5.7 (left) and Table 5.3.

It was subsequently realized that OH groups are unavoidable in the APCVD SiO_{2-x} films deposited at room temperature, irrespective of the MR value. Furthermore, the Si-OH group concentration is proportional to the film thickness. In chapter 6 the OH impurities are proven to induce deficiencies as to solar cell applications. Consequently, post-deposition treatment such as calcination of the deposited films is essential to be investigated (shown later in section 5.6).

In terms of APCVD SiO_{2-x} film stoichiometry, it was revealed that the Si-O peaks slightly shift in the direction of low wavenumber as the MR increases, shown in Figure 5.7 (right). The shift of the Si-O band frequency to a lower wavenumber is attributed to the increase of Si atoms in the oxide film [25]. This, in turn, leads to a slight improvement in the oxide film stoichiometry. Figure 5.7 (left) reveals that the Si-OH bands shift towards lower wavenumbers as their relevant MR increases, which indicates that Si atom concentration is higher in the oxide film matter. This is due to the fact that Si-O bonds are weaker than those of Si-H, H-OH and Si-OH and the Si atomic mass (28.08 u) is higher than that of H (1 u). However, it is inefficient to progress the film stoichiometry by means of setting an extremely high MR (e.g up to 10), as the parasitic hydroxyl groups are consequently enriched in the oxide film, as well. This might lead to a stoichiometric SiO_2 clusters incorporated in a void-rich SiO_x film. This can be investigated by defining the APCVD SiO_{2-x} optical constants. A further discussion in this regard is provided in section 5.5.1.

5.4.2 Study of the APCVD SiO_{2-x} film composition as a function of the deposition temperature

It was previously shown that the T_s significantly influences the APCVD SiO_{2-x} DR because of its proportional impact on the MR value. MR, in turn, was proven to influence the film chemical composition in the precursor section. This suggests that T_s possesses an impact on the film chemical composition, as well. In this section the effect of T_s on the APCVD SiO_{2-x} film chemical bonds and stoichiometry is studied by means of FTIR inspection.

For that matter, one set of the samples investigated in section 5.3.3 was utilized as an experiment material. The samples chosen were deposited at MR= 4.5 at a temperature range from 5 to 50 °C with an interval of 5 °C. One sample of a shiny etched 2×2 cm^2 FZ Si wafer without coating was used as a background reference for comparison with the coated ones during FTIR measurements. The FTIR study of the deposited films was carried out with VERTEX 80v instrument from Bruker. The samples were measured using the room-temperature detector and the chamber was under low pressure of about 3 - 4 kPa. The spectral wavenumber was in the range from 400 to 4400 cm^{-1} with a resolution of 8 cm^{-1}. The individual measurement duration was set to 60s at an iteration of 10. The baseline corrections of the spectra, subtracting reference data from those of the oxide film, and peak pickings were executed with Origin 2020 software.

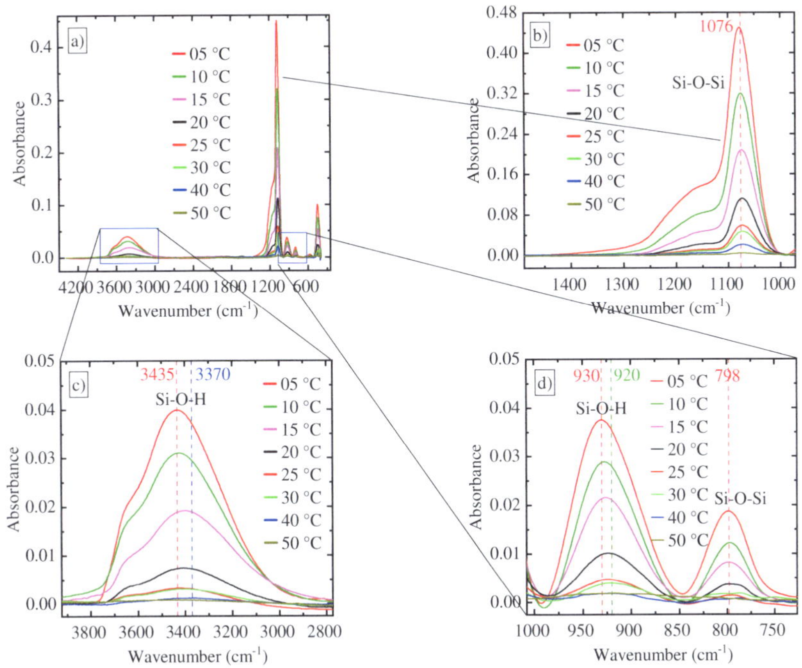

Figure 5.8: a) A graph showing the FTIR spectra of 8 APCVD SiO_{2-x} films prepared at a T_s range from 5 to 50 °C, at MR of 4.5. Enlarged portions of the spectra showing the absorption bands at **b)** a wavenumber range of 900 – 1450 cm^{-1}, **c)** a wavenumber range of 700 – 1000 cm^{-1}, and **d)** a wavenumber range of 2800 – 3900 cm^{-1}

The FTIR measurements resulted in the spectra shown in Figure 5.8. For clarity, the spectra were magnified to be legible enough and divided to three portions corresponding to three different absorption bands at various wavenumber ranges. Figure 5.8 revealed two distinctive chemical bonds in the oxide film: Si-O and Si-OH. As to the bond densities, it is shown that they are inversely proportional to the T_s. This bespeaks the decrease of the whole bond concentrations, which is, in turn, an evidence of the decrease in the APCVD SiO_{2-x} film thickness as the T_s increases. As to Si-O bonds it is demonstrated that their corresponding peaks were symmetrically aligned at a constant wave number without any shift. This holds true for their both bands positioned at 789 and 1076 cm^{-1}. However, these peak positions are still far from those corresponding to the stoichiometric oxide. It was, therefore, concluded that the deposition temperature does not play any role in the SiO_{2-x} film stoichiometry. On the other hand, the bands corresponding to Si-OH groups are shown to shift to a lower wavenumber as the T_s increases. This shift occurs from 3435 to 3370 cm^{-1}, as well as, from 930 to 920 cm^{-1}, each at T_s range from 5 to 40. Consequently, it can be concluded that as the T_s increases, the Si-OH bonds are strengthened and enriched with Si atoms, yet the OH groups are still present in the APCVD SiO_{2-x} films. In the following chapter these finding are to be confirmed by means of investigating the optical constants of the oxide films.

5.5 Impacts of the APCVD boundary conditions on SiO_{2-x} optical properties

5.5.1 Impact of the $SiCl_4$-to-H_2O MR on *n* & *k*

A set of chemically etched 2×2 cm^2 wide FZ Si wafers were used after RCA cleaning. The wafers were coated with APCVD SiO_{2-x} films, each at a various $SiCl_4$-to-H_2O MR. The T_{APCVD} was 25 °C all over the experiments. The as-deposited films were characterized by ellipsometry measurements as to their refractive indices (*n*) and extinction attenuation coefficients (*k*). The tool employed was a Woollam M-1600 spectral ellipsometer. The coated samples were measured at one angle of 70° and the relevant wavelength range was from 250 to 1000 nm. Cauchy model was chosen to describe the indices spectrum considering the exclusion of the native SiO_2 thin layer primarily present beneath the APCVD SiO_{2-x} one. The results were subsequently compared with the thermally prepared SiO_2 films [26]. The precision of the measured (n) values was in the range of 0.002.

Figure 5.9 (left) reveals that the *n* value (detected at the visible wavelength λ_{vis} = 632.8 nm) increases proportionally with the set MR. However, the greatest *n* obtained in this experiment is smaller than that of the thermal SiO_2 films, even when the APCVD SiO_{2-x} is prepared at extremely high MR of 11. As shown in Figure 5.9 (right), at MRs lower than 0.5 (e.g. MR = 0.3) the obtained *n* is quite low (*n* = 1.22). An abrupt increase in *n* value is induced at elevated MR to 0.5. As the MR continues to increase to MR = 1, *n* value increases with a lower interval. At MRs higher than 1, an extremely slight change in the *n* value is obtained (*n* = 1.45 ±0.002). This suggests that at MRs lower than 0.5 the deposited oxide films are porous/void rich, as well as, non-stoichiometric since *n* = 1.46 for the stoichiometric SiO_2.

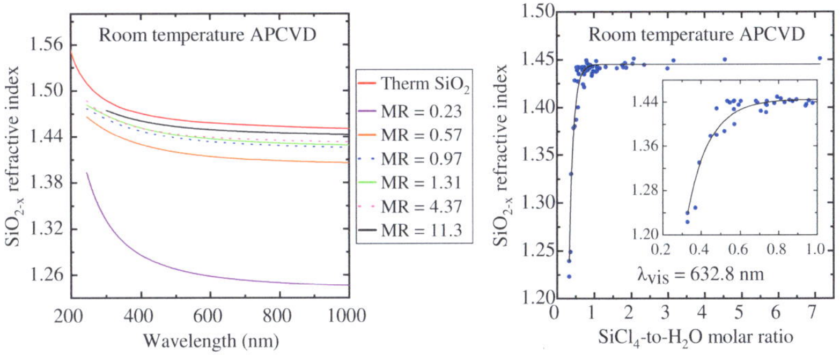

Figure 5.9: Left: Measured spectral refractive index spectrum of APCVD SiO_{2-x} films prepared at elevated $SiCl_4$-to-H_2O MRs at room temperature in comparison with a thermally prepared SiO_2 film. **Right:** Measured refractive index change of APCVD SiO_{2-x} films as a function of the MR. *n* is detected at the visible wavelength λ_{vis} = 632.8 nm. The inset shows a magnified part of the curve corresponding to the MR range from 0.3 to 1.

These findings are attributed to the chemical composition of the deposited oxide film; e.g. at low MR of 0.3 the Si content in the oxide film is low and the film is silicon poor SiO_x. This is due to the low $SiCl_4$ concentration in the injected gas mixture. As the MR exceeds 0.5, the Si content in the film evolves to values near that of the stoichiometric one. This is consistent with the findings in sections 5.3.2 and 5. 3. 3. The stabilization of *n* at MRs higher than one is unclear at this phase and further investigation are to be conducted later in this chapter.

An inspection of *k* values linked to APCVD SiO_{2-x} films deposited at elevated MR is shown in Figure 5.10. Unlike the *k* value of the thermal SiO_2 film, which is equal to zero from 250 to 1000 nm [26], it is exhibited that the *k* of the APCVD SiO_{2-x} films is rather larger than zero at both the ultraviolet (UV) wavelength (100 – 400 nm) and the visible one (400 – 700 nm). Moreover, it is seen that as the MR increases, the *k* spectrum slightly drops approaching zero, yet not completely, since even at high MRs (e.g. ≥ 4.48) there is a considerable *k* value restricted in the UV range. Consequently, it is indicated that a light absorption occurs within the oxide film throughout the UV wavelength.

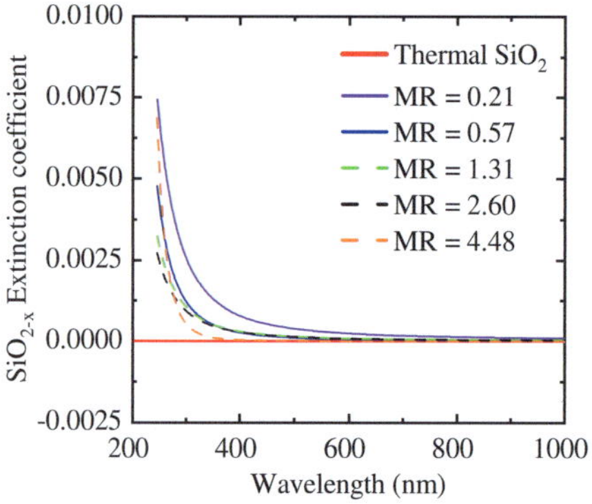

Figure 5.10: Measured spectral extinction coefficient of APCVD SiO_{2-x} films prepared at elevated MRs at room temperature in comparison with the extinction coefficient of thermal SiO_2

It is expected that due to void integration and OH impurities, e.g. silanol groups, present in the APCVD SiO_{2-x} film, a UV light scattering takes place, which exacerbates along with the decrease in MR (e.g. at 0.21). Besides, the silanol groups, characterized by a bandgap lower than that of stoichiometric SiO_2 (e.g. pure silica), cause shifting of the UV edges towards the visible wavelength. This is another indication of the presence of OH impurities in the film, which varies according to the H_2O content in the gas mixture. See the curve difference between MR 0.21 and 4.48.

5.5.2 Impact of the deposition temperature on *n* & *k*

FZ Si wafers with 2×2 cm^2 area were exploited after chemically etch and RCA clean. The samples, to be coated with APCVD SiO_{2-x} films, were divided to 5 groups. Each sample group was undergone an oxide deposition at a constant MR. The samples in the single group were coated, each at an elevated temperature. The temperature range was from 0 to 50 °C with a logarithmic stepwise. *n* and *k* values of the as deposited oxide films were measured with Woollam M-1600 spectral ellipsometer at one angle of 70° and a wavelength range was from 250 to 1000 nm. It is worth noting that the characterization of the oxide films deposited at T_s lower than 10 °C was difficult with the ellipsometry. This is due to the fact that the structure of the oxide films deposited at low temperatures is quite different from those deposited at the room one or above. It is proven below that the films deposited at low T_s are characterized by high pores and granular structure leading to a light scattering when exposed to e.g. ellipsometer light beam. Therefore, the samples deposited at temperatures between 0 and 10 °C were excluded from the ellipsometry characterization. They were, alternatively, characterized by SEM.

Figure 5.11 (left) reveals that the oxide films deposited at T_s below 12 °C are characterized by low *n* value (down to 1. 21), irrespective of the set MR value. Figure 5.11 (right) demonstrates that film deposited at T_s of 11.5 °C features a relatively high *k* value, not only in the UV range but also all along the inspected wavelength. This is an indication of light absorption due to a porous structure of the oxide film, wherein light scattering takes place. At lower T_s, at which the DR increases, it is shown that the film porosity increased. Consequently, extremely porous films with low density and rough surface morphology were obtained, demonstrated in Figure 5.12. In section 5.4.2 the APCVD at low T_s was proven to cause incorporation of large volume of OH groups, including H_2O, into the film. The same effect can appear when cooling the substrate, as such a maximum volume of H_2O molecules are adsorbed upon the Si surface e.g. due to extremely fine H_2O droplet condensation. This results in a thick film of Si-poor SiO_x, hence a low-density film structure. H_2O molecules and air are supposedly the possible candidates to occupy the film voids, either during deposition or after storing, which propose inducing partial light absorption. However, further investigation is required to prove this effect.

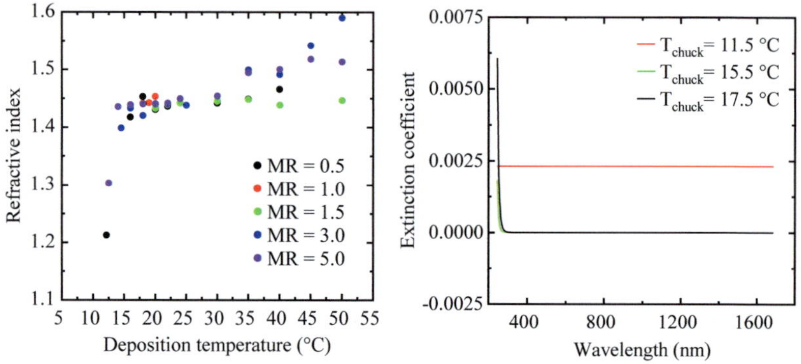

Figure 5.11: Left: A graph exhibiting the refractive indices of 5 groups (dedicated by a specific MR) of APCVD SiO_{2-x} films prepared at elevated substrate temperatures from 12 to 50 °C. **Right:** A graph exhibiting the extinction coefficients of APCVD SiO_{2-x} prepared at various temperatures up to the room one

At T_s between 14 and 35 °C the n value was increased between 1.4 and 1.45. Surprisingly, at T_s range from 35 to 50 °C the n value increases drastically to almost 1.59 at high MRs of 3 and 5. Meanwhile, the films deposited at MRs of 1.5 or below, feature almost a stable n value irrespective of the tuned T_s. This unnecessarily reflects a high oxide film stoichiometry e.g. when $n = 1.59$. It was found in section 5.4.2 that the APCVD SiO_{2-x} film stoichiometry is unchangeable, irrespective of the deposition temperature (in the studied range from 5 to 50 °C). Moreover, the SiO_{2-x} films include a considerable amount of hydroxyl groups and the vibration of Si-O bond frequency was at 1076 cm^{-1}, which is higher than that of the stoichiometric SiO_2 (1067 cm^{-1}). Another proof of the non-stoichiometry of high-n APCVD SiO_{2-x} is shown in chapter 6, wherein films featuring ($n = 1.66$) as-deposited at MRs of 10 to 20 permitted texturization of the underlying Si surface.

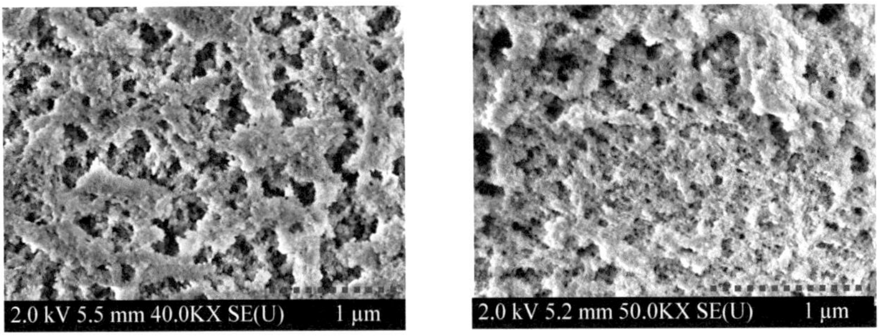

Figure 5.12: SEM images of SiO_{2-x} films deposited onto shiny etched FZ Si wafers at 5 °C (left) and 14 °C (right) exhibiting a granular, rough and porous structure

5.6 Impacts of post-deposition annealing on the APCVD SiO_{2-x} integrity

On the bases of the above, it was found that the newly developed APCVD SiO_{2-x} films can range from far non-stoichiometric to near-stoichiometric. However, a completely stoichiometric film featuring O:Si atomic ration of 2:1 has been so far difficult without annealing. The OH group content was unavoidable despite the high MR adopted and the resulting n and k values. This gave rise to the thermal treatment of the post-deposited films as a solution to avoid the OH groups' incorporation. In the following sections the impact of film calcination on the chemical and optical properties of the APCVD SiO_{2-x} was investigated.

5.6.1 Annealing impact on the APCVD SiO_{2-x} chemical composition

In order to investigate the annealing impact on the chemical composition of the APCVD SiO_{2-x} films, FTIR study was carried out. For that, 4 sample groups made of 2×2 cm^2 wide 200 μm thick shiny etched FZ Si wafers were exploited. Each group contained 6 samples. All sample groups were cut from the same 6-inch wafer precursor after RCA clean. Each sample group was coated with an APCVD SiO_{2-x} film during one run so that the same film thickness was obtained all over the 6 samples. The deposition conditions were as follows: MR = 1.74, T_{APCVD} = 32.6 °C, and a deposition duration of 10 min. An additional sample from the same Si wafer precursor was left without coating to be used as a reference background for the FTIR measurements. Subsequently, 5 samples from each group were annealed in ambient air for 5 min, each at a various temperature using a hotplate. The temperature range was from 200 to 600 °C with a stepwise of 100 °C. One sample from each group was left without annealing so as to be used comparison reference. Ultimately, each sample group was FTIR inspected all at once using the sample holder of the wheel kind. This wheel holder (available in VERTEX 80v FTIR spectrometer version provided by Bruker) permits mounting 7 samples to be measured successively without the need to break the vacuum for individual measurements. The baseline correction of the resulting spectra and subtraction of reference data from those of the APCVD SiO_{2-x} film, and peak determinations were conducted using Origin 2020 software. The aforementioned process was performed for all 4 sample groups. The resulting FTIR spectra were identical, thus we chose one of the group measurements for result visualizations.

It was found that the calcination process flattens the curve of impurity presence in the APCVD SiO_{2-x} films, irrespective of the deposition boundary conditions. Demonstrated in Figure 5.13 (right), the Si-OH transmittance bands at 916 cm^{-1} were completely removed at an annealing temperature (T_a) of ≥ 500 °C for 5 min. It is to note that at temperatures ≤ 200 °C a minor influence on the film composition was obtained. Meanwhile, the effective temperature range was from 300 to 600 °C. Looking at the bands' shift, it is observed that the Si-O stretching bands shifts from the wavenumber 1076 cm^{-1} to a lower wavenumber of 1069 cm^{-1} (near the stoichiometric) after annealing at 600 °C [16,17]. Moreover, the full width at half maximum (FWHM) (or represented by the broadness of the peak) of these peaks are higher than that of the unannealed film, hence denser Si-O bonds per oxide volume. The weak Si-O bending band, on the other hand, shifts from 796 cm^{-1} to a higher wavenumber of 812 cm^{-1} after annealing at 600°C. This indicates that more Si is incorporated in the film and the film is Si-rich silicon oxide [25]. Figure 5.13 (left) reveals that the OH group concentration in the film declines along with the raise of calcination temperature. However, these impurities were not eliminated in the film completely. Furthermore, residual H-O-H bonds or free H_2O molecules remained in the film, as well. Elimination of the H-O-H bonds in the oxide film was, therefore, further strived. Calcination of APCVD SiO_{2-x} films at 600 °C was conducted for three gradual durations: 5, 10 and 15 min, and the samples were subsequently compared with unannealed one using FTIR investigation.

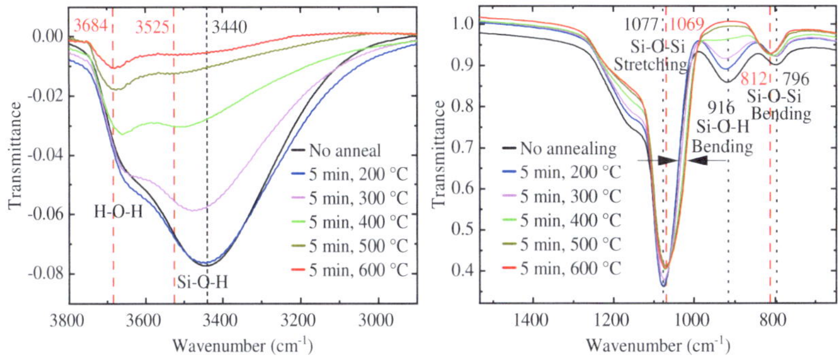

Figure 5.13: Graphs depicting enlarged and baseline corrected FTIR spectra of APCVD SiO_{2-x} films without and with calcination for 5 min at various temperature range from 200 to 600 °C, as well as, their respective absorption band at **(left)** the wavenumber 3000 – 4000 cm^{-1} and **(right)** the wavenumber 600 – 1400 cm^{-1}

Figure 5.14 demonstrates that the longer the calcination lasts, the lower the intensity of Si-OH and H-OH bonds in the oxide volume – e.g. the annealing duration of 15 min at 600°C removed nearly all the Si-OH bonds and reduced the free H-OH bonds to a further extent. A negligible portion of H-OH bonds was shown to remain unchanged in the oxide film irrespective of the calcination time. Further investigation of the activation energy of calcination is still required. In chapter 6 this extremely small amount of H-OH bonds is proven to possess no impact on the film applications in the solar cell manufacture. Surprisingly, the oxide film annealed at 200 °C was shown to improve the film resilience so that it stands electroplating currents used for metallizing SHJ solar cells (the reader is referred to chapter 6 for detailed information).

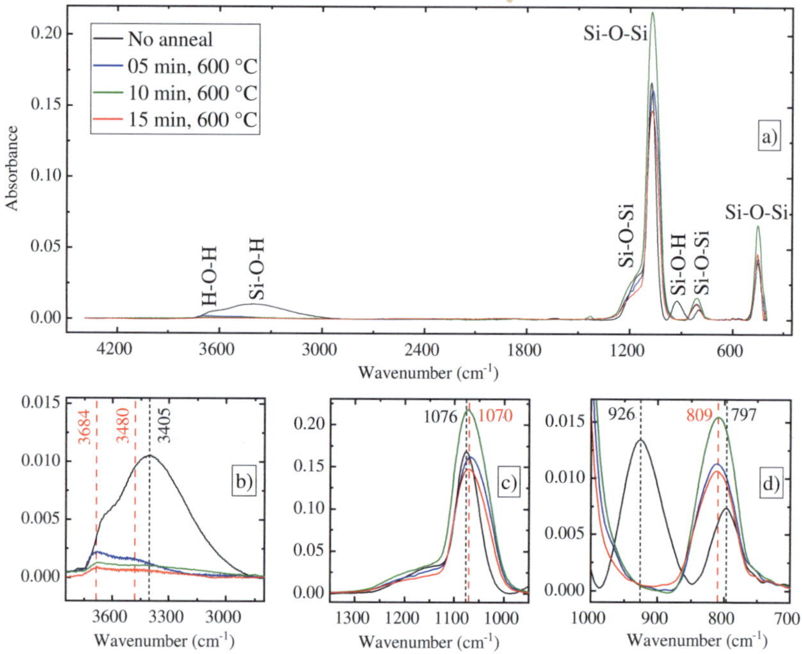

Figure 5.14: a) A graph showing FTIR spectra of APCVD SiO_{2-x} films annealed at 600 °C for three various durations: 5, 10, and 15 min, in comparison with an unannealed sample. Enlarged portions of the spectra shown in a) at the wavenumbers: **b)** 2850 – 3750 cm^{-1}, **c)** 1000 – 1300 cm^{-1} and **d)** 700 – 1000 cm^{-1}

It is important to note that the calcination process induced a shrinkage of the APCVD SiO_{2-x} film, which is remarked by the reduction of the absorption band intensities (or gauged by lowering the peak height) along with the annealing temperature and time, see Figures: 5.13 and 5.14. This is attributed to the release of OH groups from the films. The association of film shrinkage with calcination is an evidence of the low density of the as-deposited APCVD SiO_{2-x} film. However, it is inadequate to completely associate the shrinkage with densification of the film after annealing, as shown in the next section. The band corresponding to the samples annealed for 10 min (Figure 5.14) is abnormally higher than the others (accidently occurred during deposition) because the related as-deposited film thickness was larger than the other inspected films, thus more bonds are detected by the IR spectra.

It is worth noting EDX measurements revealed that annealing the coated Ge samples prepared in section 5.2 at 300°C for 1 min progressed the Si:O atomic ratio from 1:1.8 to1:1.93.

5.6.2 Annealing impact on the APCVD SiO_{2-x} etch rate

The effect of calcination on film resistivity against etchants was traced. 2×2 cm^2 wide shiny etched and RCA-cleaned FZ Si wafers were coated with about 160 nm thick APCVD SiO_{2-x} films. The oxide coated samples were divided into three sets; the first, second and third sets were annealed for 5, 10 and 15 min, respectively. The samples of each set were annealed at elevated temperatures from 100 to 600 °C, each by 100 °C stepwise. The annealing was executed in ambient air on a hotplate. For each set, one sample was left without annealing so as to be compared with the annealed samples. The unannealed and annealed samples were subjected to etching in 1 wt% hydrofluoric acid (HF) solution for 10s, after measuring the coating film thickness by means of ellipsometry. Subsequently, the film thickness was measured after etching with ellipsometry, and thus the wet etch rate (ER) was, ultimately, calculated. Woollam M-1600 spectral ellipsometer was employed using one angle of 70 and the inspected wavelength was from 250 to 1000 nm.

Figure 5.15 revealed that the APCVD SiO_{2-x} ER gradually decreases along with the increase of T_a to a temperature between 300 and 400 °C. Wet ER of the unannealed films is measured between 1.63 to 1.68 nm/s, whereas it is reduced to 0.1 nm/s of the annealed samples at 300 °C for 15 min. Surprisingly, wet ER increases again for samples annealed at the temperatures 500 and 600 °C. The high T_a, ranging from 500 to 600 °C, appears to not induce proper stability of the APCVD SiO_{2-x} films against wet ER. However, due to the high error bars of the plot, there might be a decrease in the wet ER and therefore further investigations are recommended. If wet APCVD SiO_{2-x} ER increase cannot be explained by film densification, there are other reasons overshadowing the calcination effect. The following explanation of the 500 –

600 °C effect could be an option. A. Theil *et al.* [16] found that the Si–O stretching bands positioned at low wave numbers, corresponding to Si-rich SiO_x, are combined with strained bonds, or so-called dangling bonds. This, in turn, goes hand in hand with a tensile stress characterizing the oxide film.

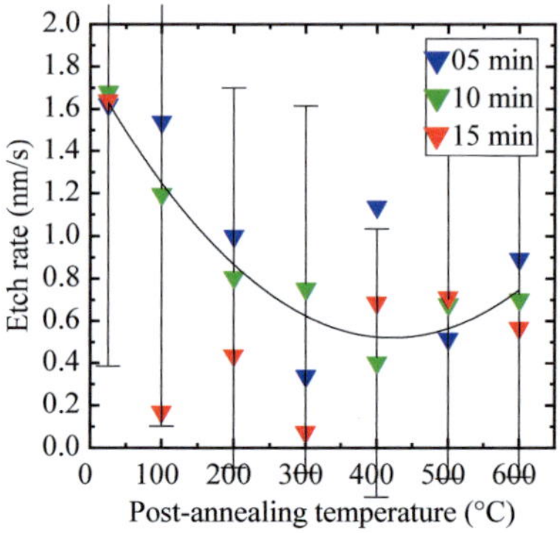

Figure 5.15: A graph demonstrating the etch rate of APCVD SiO_{2-x} films with and without annealing for various durations as a function of the annealing temperature. The continuous line is a guide for the eye. The error bars show large standard deviations range of 0.6 – 1.3 (calculated for the square root values).

In this case of research, the annealing process, at temperatures above 500 °C, results in a shift of the stretching Si–O band toward lower wavenumbers, in combination with reduction in bond intensity. Consequently, the APCVD SiO_{2-x} converts to Si-rich kind (shown in Figures 5.13 and 5.14). This film characteristic is similar to that found by H. G. Tompkins and A. Theil. This suggests that the reason of the APCVD SiO_{2-x} film deficiencies is the mechanical stress developed due to disorder in the Si–O networks, which appears at high calcination temperatures. The high calcination temperature results in releasing nearly the whole hydroxyl groups leaving Si–O dangling bonds, hence, vacancies or perhaps nano-cracks in the oxide film structure. This in turn causes sub-densification of the film, hence a tensile stress. Since the presence of Si–H bonds in these films is below the detection limit and uncounted, the vacancies, after annealing, remain empty which would promote fast ER in wet HF, and even worse, possible poor electrical properties. This suggests that the newly developed APCVD SiO_{2-x} films do not play a passivation role for the Si surface. However, it is worth investigating the possibilities of extending these films by the function of interface passivation and dopant sources for Si. Further investigations to trace the film behavior after storage, e.g. in ambient air and humid environment, as well as, the film electrical properties are still worthy.

Another evidence of the aforementioned film deficiency at high calcination temperature is displayed on the basis of optical constants in the following section.

5.6.3 Annealing impact on the APCVD SiO_{2-x} optical properties

Annealing the APCVD SiO_{2-x} films was found to diminish the film volume in combination with inadequate densification. In order to evaluate this effect, a systematic study of thickness change and optical constants of the newly developed oxide films was conducted. The APCVD SiO_{2-x} coated samples previously used for etch rate evaluation were inspected before annealing, after annealing, and after etching in 1 wet% HF solution. Ellipsometry was used to measure the thickness (t_{SiO2-x}) and *n* value of the oxide films. Subsequently the reduction in the APCVD SiO_{2-x} film thickness (Δt_{SiO2-x}) after annealing was calculated and plotted in Figure 5.16 as a function of T_a, as well as, the corresponding *n* values were summarized in Table 5.4. It is demonstrated in Figure 5.16 that the Δt_{SiO2-x} increases gradually with the increase of T_a, which is consistent with the findings in section 5.6.1. It is, as well, demonstrated that the annealing duration causes a minor impact on the Δt_{SiO2-x} at $T_a \leq 300$ °C. On the other hand, at T_a between 400 and 600 °C the duration impact is more obvious, wherein the longer the annealing duration, the larger the Δt_{SiO2-x}.

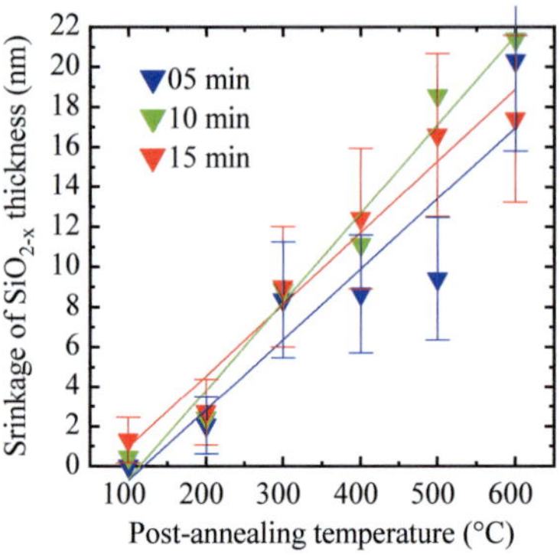

Figure 5.16: A graph elucidating the shrinkage of the APCVD SiO_{2-x} film thickness after annealing for three various durations: 5, 10 and 15 min as a function of the annealing temperature. The straight lines are guides for the eye. The error bars demonstrate an average standard deviation of 2.5 (calculated for the square root values).

Table 5.4: n values of the APCVD SiO_{2-x} films measured before and after annealing at elevated Ta for two different durations: 5 and 15 min. The first two decimals are the considerable ones.

n value	5 min annealing			15 min annealing		
T_a	As-deposited	After anneal	After wet HF	As-deposited	After anneal	After wet HF
100 °C	1.441	1.440	1.440	1.443	1.442	1.441
200 °C	1.442	1.443	1.440	1.441	1.437	1.439
300 °C	1.441	1.442	1.445	1.443	1.441	1.442
400 °C	1.443	1.440	1.445	1.441	1.437	1.444
500 °C	1.444	1.446	1.442	1.439	1.450	1.442
600 °C	1.443	1.427	1.436	1.443	1.428	1.433

At first glance, it is presumed that the more the APCVD SiO_{2-x} film shrinks under calcination, the more compact it has, and hence it possesses higher ρ_{SiO2-x} and *n* values. This, in turn, translates to tensile stress built up in the oxide film. This assumption is in accordance with the findings in section 5.6.2, wherein films annealed e.g. at 600 °C display higher ERs, which is attributed to vacancies integrated in the film, reflecting low SiO_{2-x} density (ρ_{SiO2-x}). Despite it is frequently accepted that SiO_2 film density (ρ_{SiO2}) and its *n* value are proportionally linearly correlated [27], there has been, thus far, no proper account proving similar effect as to APCVD oxides prepared with the hydrolysis of $SiCl_4$. In the following discussion the relationship between *n* and ρ_{SiO2-x} values is to be realized with the assistance of *n* value after wet etching in HF solution.

Table 5.4 exhibits that at $T_a \leq 400$ °C there is a slight difference in *n* values before and after annealing; nevertheless, it is still lower after annealing in most of the experiments. On the other hand, *n* values considerably differ before and after annealing at $T_a = 500$ and 600 °C. Annealing the oxide film at 500 °C is shown to anomalously increase the *n* value with about $\Delta n = 0.01$. In contrast, annealing at 600 °C for 5 and 15 min is shown to again decrease the *n* value, each with about $\Delta n = -0.016$. The terms minus and plus are used for distinguishing between the increase and decrease in *n* values.

The latest observation evidences that in spite of film shrinkage, the ρ_{SiO2-x} decreases after annealing at 600 °C, which agree with the finding in the previous sections. The high *n* values measured before annealing are ascribed to SiO_{2-x} enriched with Si atoms. However, this oxide features low ρ_{SiO2-x} due to impurities and voids, opposing what is known in ref 24. That is why *n* values after etching are larger than those before, which proposes Si-related clusters engaged in the film. After annealing, the hydroxyl/water impurities are diffused through the film bulk and released to the ambience. Consequently, tensile stress is built up at high T_a such as 600°C inducing enlargement of the original voids, hence lower *n* values, illustrated in Figure 5.17. The tensile stress in the oxide film causes a compressive stress in the top Si substrate boosting the potential of cracks in the oxide coating or in severe cases film delamination. The response of the film at T_a of 500 °C requires more investigations. Further inspection to prove this analysis is recommended by means of other characterization methods than FTIR and ellipsometry.

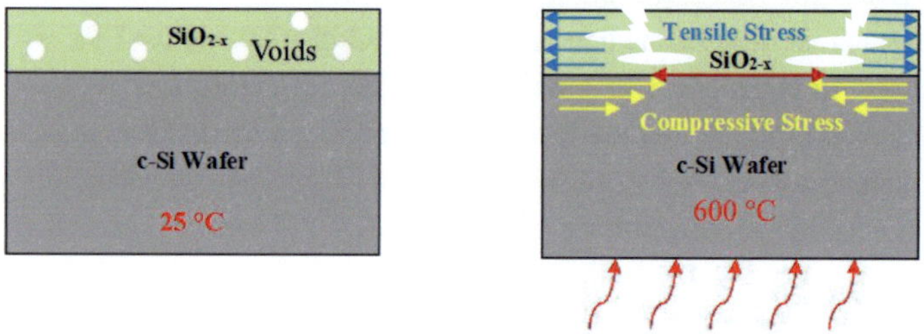

Figure 5.17: Schematic drawing of **left:** a porous APCVD SiO_{2-x} film right after deposition and **right:** the tensile and compressive stress between the porous APCVD SiO_{2-x} film shown left and the underlying Si wafer surface after annealing at e.g. 600 °C

5.7 Investigation of structural defects in the APCVD SiO_{2-x} films

The APCVD SiO_{2-x} films inferiority, specifically when it comes to solar cell applications, can be caused by cracks, porous structure/voids, low break through voltage, non-uniform thickness, high impurities, instability against alloy diffusion, dust formation, etc. In the following sections, an investigation is carried out so as to set up the appropriate boundary conditions and post-deposition treatment that would ultimately lead to a reliable film integrity.

5.7.1 Impact of deposition boundary conditions on the film structure

In this experiment APCVD SiO_2 films were prepared at severe boundary conditions such as high DR = 140 nm/min and AH_m = 28 g/m³, corresponding to low $SiCl_4$-to-H_2O MR = 0.3, and relatively low T_s of 17.5 °C. The resulting film thickness was in the range from 600 to 900 nm. The effect of these conditions on the film structure was investigated before and after annealing at T_a = 300 – 400 °C for 1 min on a hotplate in ambient air. In order to examine the film characteristics on top of various surface-structures, they were deposited onto two different sets of substrates, which are to be exploited later in the fabrication of various solar cell kinds:

i) shiny etched 2×2 cm^2 wide FZ Si wafers after RCA clean,
ii) alkaline-textured and ITO coated 2×2 cm^2 wide FZ Si wafers.

The films underwent an SEM inspection before and after annealing using Bruker Quantax 400 in a Zeiss Auriga 60 tool. The SEM acceleration voltage was adjusted at 5 kV and the working distance to the surface was in the range from 4.9 to 5.2 mm.

Figures 5.18a) and b) demonstrate that the oxide films as-deposited onto both substrates feature a granular surface and a porous structure (similar to the so-called bread-loafing). This is ascribed to the evaporation of H_2O molecules from the film inducing shrinkage and hence a tensile stress in the film. The film structures shown in Figure 5.18 indicate a low ρ_{SiO2-x}. It is worth mentioning that measuring the *n* value of the film shown in Figure 5.18 was not possible with ellipsometry because of light scattering within the film. Figures 5.18c) and d) demonstrate the oxide films shown in a) and b) after annealing at 400 °C. It is shown that both films crack and delaminate, irrespective of the substrate surface morphology. This observation confirms the findings in section 5.6.3, as such the oxide film with poor optical properties is more susceptible to damage by tensile stress. Depending on the porosity degree and the volume of hydroxyl groups in the as-deposited film, the tensile stress developed from annealing induces breaking the film, propagating the existing cracks, and pealing-off the film from the substrate surface. In chapter 6 it is shown that films deposited at lower MR abruptly crack and fall-off when annealing even at 100 °C.

5.7.2 Impact of the film thickness and substrate surface structure on the film durability

In this section deposition boundary conditions, known to produce APCVD SiO_{2-x} film properties aligning those of the thermal SiO_2, were adopted. $SiCl_4$-to-H_2O MR adjusted at 1, meanwhile, T_{APCVD} and T_s were maintained at room temperature all over the experiments. Two sample sets were exploited in order to investigate possible structural defects in the APCVD SiO_{2-x} films and their integrity.

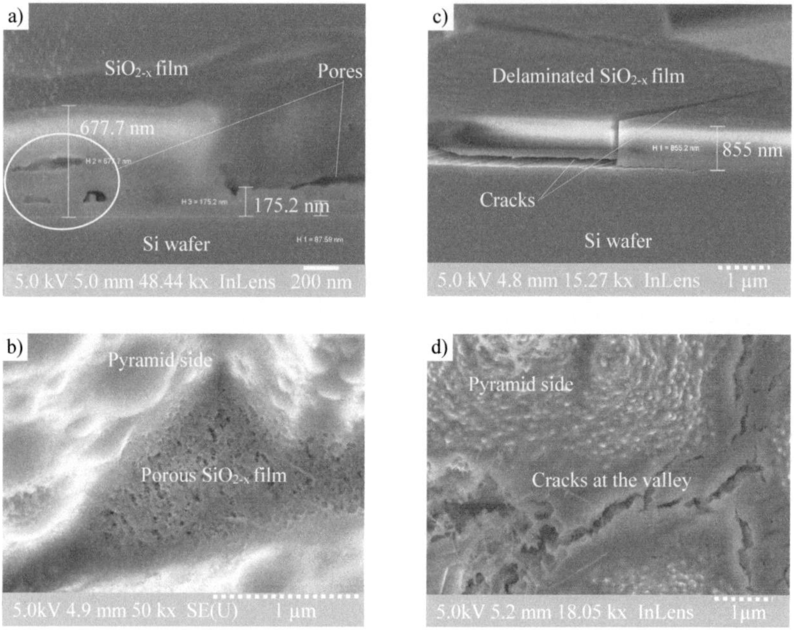

Figure 5.18: a) and b): SEM images illustrating inferiorly porous APCVD SiO_{2-x} films prepared at a low MR of 0.3 and T_a of 17.5 °C, and applied upon shiny etched FZ Si wafer (**a:** cross section) and alkaline textured and ITO coated Si wafers (**b:** tilted top view), respectively. **c) and d):** SEM images illustrating the APCVD SiO_{2-x} films shown in a) and b) after annealing at 300 – 400 °C for 1 min, respectively. Delamination and cracks are formed in the films coating the planar Si surface, as well as, the pyramid sides and valleys.

In the first set, an experiment was performed to investigate the influence of the film thickness on its microstructure and the interface with the Si wafer surface characterized by almost atomic planarity. For that, a range of 80 to 800 nm thick APCVD SiO_{2-x} films were deposited onto chemically etched FZ Si wafers with an area of 2×2 cm^2 after RCA cleaning. In the second sample set experiments were carried out to check the film durability and conformity upon different surface structures, hence its compatibility with various PV applications. Three sets of 2×2 cm² large samples to be investigated were:

i) alkaline textured mono-crystalline Si wafers
ii) alkaline textured and SiNx coated mono-crystalline Si wafers
iii) alkaline textured and ITO coated mono-crystalline Si wafers

The three sample sets were coated with about 115 - 130 nm thick APCVD SiO_{2-x} films. Owing to the fact that the newly developed APCVD reactor permits mounting 6-inch large substrates, one wafer from each sample set was placed in the APCVD apparatus and all three wafers were coated simultaneously during one run. The samples were subsequently inspected with SEM adjusting the acceleration voltage between 5 and 6 kV and the working distance from 4.8 to 5.2 mm. The film thickness and optical constants were measured with a Woollam M-2000 spectral ellipsometer. The measurements were conducted at three various angles of 65, 70 and 75° and a relevant wavelength range from 300 to 900 nm. These experiments and their associating investigations are published in the "Thin Solid Films" journal [28].

The experiment results are demonstrated in Figure 5.19. In the first experiment set, where the change in film thickness was traced, dense APCVD SiO_{2-x} films without cracks or pores are shown onto the wafers irrespective of their thicknesses. The Si surface is coated with the as-deposited SiO_{2-x} intimately without gaps. Moreover, as shown in Figure 5.19a, b, c, the top surface of the SiO_{2-x} films is shown to be smooth without adhered SiO_{2-x} contaminants and without oxide dust lying on the surface, opposing conventional APCVD techniques [8,9]. As shown in Figure 5.19d, e, and f, the APCVD SiO_{2-x} films were deposited conformally on the substrates irrespective of their surface-morphology and underlying materials. The random pyramids of i) the alkaline textured, ii) the textured and SiN_x coated iii) and the textured and ITO coated samples were conformally and uniformly coated with the APCVD SiO_{2-x} film. No voids or cracks are observed in the SiO_{2-x} film, as compared to the ITO layer underneath, shown in Figure 5.19f.

Figure 5.19: SEM images of APCVD SiO_{2-x} films deposited on various surface structures of FZ Si wafers exhibiting **a)** top view with 10° tilted angle of 220 nm thick film **b) and c)** 10° tilted top view and cross section of 595 nm thick film, respectively. Images **d** to **f** show cross section of about 115-130 nm thick APCVD SiO_x films grown conformally on **d)** an alkaline textured mono-crystalline FZ Si wafer, **e)** an alkaline textured and SiN_x coated mono-crystalline Si wafer, **f)** and an alkaline textured and ITO coated mono-crystalline Si wafer, respectively. The bright line between the Si and SiO_x in d) is attributed to charging effect by the SEM beam. The yellow dashed line in e) is to a guide for the eye to distinguish between the SiO_{2-x} and SiN_x layers.

In terms of stability after annealing, the post-deposited SiO_{2-x} films shown in Figure 5.19, numerous tests were executed at different temperatures and durations, depending on the future application of the relevant film. The experiments were conducted using a hot plate and a ceramic shielded furnace. The annealing experiments proved the followings:

- The films deposited on textured and ITO coated Si wafers withstand annealing temperature of 250 °C for 2 hours. The films remain compact and intimate with the ITO surface.
- The films deposited on textured and SiN_x coated Si wafers are stable against annealing at 870 °C for 4 sec, as well as, at 400 °C for 30 min. No peel-off or cracks were shown in the film.

- The films deposited on shiny etched FZ wafers stand annealing in a furnace at 900 °C for 7 hours. The film structure and adherence to the wafer remained the same as before annealing.

It is to note that the APCVD SiO_{2-x} films mildly shrink after annealing (see Figure 5.16), however their performance is proven to suit all investigated applications presented in chapter 6. This is an indication of resilience improvement of the APCVD SiO_{2-x} films after annealing and nonviolent impact of the mechanical stress created during annealing. It is to note, as well, that the annealing durations used for PV applications are way much less than the examined ones in this experiment.

5.8 Summary

An analytical study of the APCVD SiO_{2-x} film growth rate and properties was carried out. The nature of the oxide film deposited under atmospheric conditions utilizing $SiCl_4$ and H_2O chemical precursors was inspected. It was shown that the films are silicon oxide characterized by stoichiometry lower than that of the thermal SiO_2 films. The Si:O atomic ratio of the film was measured to be 1:1.8 without calcination. However, after calcination the Si:O atomic ratio was measured to be at least 1:1.93. The as-deposited oxide film, was therefore considered near stoichiometric and thus named as SiO_{2-x} throughout this research. The impact of deposition boundary conditions, namely: gas flow rate, reactant concentration, molar ratio, and substrate temperature, on the film growth rate was investigated. It was shown that the film deposition rate is proportional to the flow rate of H_2O vapor, whereas it demonstrated an increase followed by decrease when it comes to $SiCl_4$ vapor flow rate. This behavior is ascribed to reduction in the $SiCl_4$ molecule concentration into the gas stream featuring flow rates higher than 1 l/min. On the other hand, the low deposition rate resulted at flow rates lower than 1 l/min is attributed to incomplete intermixing of the two gas precursors. $SiCl_4$-to-H_2O molar ration was found to be a comprehensively reliable parameter for the investigation of film deposition rate and functional properties. The film deposition rate was found to increase to its greatest value at molar ratios of 0.5 – 1.0, higher than which the deposition rate shows reduction again. The substrate temperature was found to have a significant effect on the deposition rate, as though they are inversely proportional. In terms of chemical properties of the APCVD SiO_{2-x} film, it was proven that the higher the $SiCl_4$-to-H_2O molar ratio, the lower the hydroxyl groups' concentration in the film. However, their presence is unavoidable in the as-deposited films. Furthermore, the oxide films deposited at high $SiCl_4$-to-H_2O molar ratios demonstrated a minor improvement in the stoichiometric degree. As to the impact of substrate temperature on the film chemical properties, it was found that as the annealing temperature increases, the silanol bonds are densified in the film and enriched with Si atoms. Meanwhile, the stoichiometric degree does not show any improvement independent on the substrate temperature. In terms of optical properties of the APCVD SiO_{2-x} film, it was shown that at room temperature the film refractive index increases gradually with the increase in $SiCl_4$-to-H_2O molar ratio. However, at molar ratios higher than 1, no further increase occurs in the film refractive index. The substrate temperature on the other hand was found to influence the film refractive index, thoroughly. The lower the substrate temperature below the room one, the lower the film refractive index and the higher the extinction coefficient, and vice versa, the higher the substrate temperature above the room one, the higher the film refractive index and the lower the extinction coefficient. Post-deposited annealing of the APCVD SiO_{2-x} films was found to remedy their resilience. The hydroxyl groups were minimized to a further extent at annealing temperature of 600 °C for 15 min. The film stoichiometry was improved to such an extent that it aligned that of the thermal SiO_2 films, yet not completely identical. It was also found that that the etch rate of the APCVD SiO_{2-x} film in 1wt% HF decreases with the increase of annealing temperature and time. Nevertheless, at temperatures higher than 400 °C the film etch rate exhibited increase again. It was proven that the film annealing at 600 °C causes a tensile stress in the film which, in turn, results in a high etch rate, low film density and refractive index. The film refractive index after etching was shown to increase in comparison with that after annealing. Lastly, the film structure demonstrated inferior characteristics when deposited at low temperature and $SiCl_4$-to-H_2O molar ratio and, even worse, the deposited films were endurable of annealing treatment. Contrary to that, the APCVD

SiO_{2-x} films prepared at molar ratios higher than 0.5 exhibited a sufficient durability and resilience against annealing treatment even at temperatures of 900 °C.

5.9 Bibliography

[1] M. Kawachi, Silica waveguides on silicon and their application to integrated-optic components, Opt. Quantum Electron. 22 (1990) 391–416. https://doi.org/10.1007/BF02113964.

[2] G. Lu, L.L. Tedder, G.W. Rubloff, Process sensing and metrology in gate oxide growth by rapid thermal chemical vapor deposition from SiH_4 and N_2O, J. Vac. Sci. Technol. B Microelectron. Nanom. Struct. 17 (1999) 1417–1423. https://doi.org/doi.org/10.1116/1.590770.

[3] K. Maex, M.R. Baklanov, D. Shamiryan, F. Iacopi, S.H. Brongersma, Z.S. Yanovitskaya, Low dielectric constant materials for microelectronics, J. Appl. Phys. 93 (2003) 8793–8841. https://doi.org/10.1063/1.1567460.

[4] Y. Ikeda, Y. Numasawa, M. Sakamoto, Ozone/organic-source APCVD for conformal doped oxide films, J. Electron. Mater. 19 (1990) 45–49. https://doi.org/10.1007/BF02655550.

[5] H. Nagel, A. Metz, R. Hezel, Porous SiO_2 films prepared by remote plasma-enhanced chemical vapour deposition - a novel antireflection coating technology for photovoltaic modules, Sol. Energy Mater. Sol. Cells. 65 (2001) 71–77. https://doi.org/10.1016/S0927-0248(00)00079-9.

[6] D. Adachi, T. Terashita, T. Uto, J.L. Hernández, K. Yamamoto, Effects of SiO_x barrier layer prepared by plasma-enhanced chemical vapor deposition on improvement of long-term reliability and production cost for Cu-plated amorphous Si/crystalline Si heterojunction solar cells, Sol. Energy Mater. Sol. Cells. 163 (2017) 204–209. https://doi.org/10.1016/j.solmat.2016.12.029.

[7] Krishna Seshan, Handbook of Thin-Film Deposition Processes and Techniques, William Andrew, New York, 2001.

[8] T. Fujimoto, K. Okuyama, S. Yamada, M. Adachi, Effect of cluster/particle deposition on atmospheric pressure chemical vapor deposition of SiO_2 from four gaseous organic Si-containing precursors and ozone, J. Appl. Phys. 85 (1999) 4196–4206. https://doi.org/10.1063/1.370331.

[9] J. liang Zhang, J. Li, L. ma Luo, Y. hua Wo, Microstructure and morphology of SiO_x film deposited by APCVD, J. Alloys Compd. 469 (2009) 535–538. https://doi.org/10.1016/j.jallcom.2008.02.014.

[10] J.D. Chapple-Sokol, C.J. Giunta, R.G. Gordon, A Kinetics Study of the Atmospheric Pressure CVD Reaction of Silane and Nitrous Oxide, J. Electrochem. Soc. 136 (2019) 2993–3003. https://doi.org/10.1149/1.2096390.

[11] T. Homma, R. Yamaguchi, Y. Murao, A Room Temperature Chemical Vapor Deposition SiOF Film Formation Technology for the Interlayer in Submicron Multilevel Interconnections, J. Electrochem. Soc. 140 (2019) 687–692. https://doi.org/10.1149/1.2056143.

[12] M. Matsuura, Y. Hayashide, H. Kotani, H. Abe, Film characteristics of apcvd oxide using organic silicon and ozone, Jpn. J. Appl. Phys. 30 (1991) 1530–1538. https://doi.org/10.1143/JJAP.30.1530.

[13] M. Shimbo, Thermal Stress in CVD PSG and SiO_2 Films on Silicon Substrates, J. Electrochem. Soc. 130 (1983) 135. https://doi.org/10.1149/1.2119640.

[14] A.K. Sinha, H.J. Levinstein, T.E. Smith, Thermal stresses and cracking resistance of dielectric films (SiN, Si_3N_4, and SiO_2) on Si substrates, J. Appl. Phys. 49 (1978) 2423–2426. https://doi.org/10.1063/1.325084.

[15] W. Kern, Evolution of silicon wafer cleaning technology, Proc. - Electrochem. Soc. 90 (1990) 3–19.

[16] J.A. Theil, D. V. Tsu, M.W. Watkins, S.S. Kim, G. Lucovsky, Local bonding environments of Si–OH groups in SiO_2 deposited by remote plasma-enhanced chemical vapor deposition and incorporated by postdeposition exposure to water vapor , J. Vac. Sci. Technol. A Vacuum, Surfaces, Film. 8 (1990) 1374–1381. https://doi.org/10.1116/1.576885.

[17] R. Hofman, J.G.F. Westheim, I. Pouwel, T. Fransen, P.J. Gellings, FTIR and XPS studies on corrosion-resistant SiO_2 coatings as a function of the humidity during deposition, Surf. Interface Anal. 24 (1996) 1–6. https://doi.org/10.1002/(SICI)1096-9918(199601)24:1<1::AID-SIA73>3.0.CO;2-I.

[18] G. Lucovsky, Low-temperature growth of silicon dioxide films: A study of chemical bonding by ellipsometry and infrared spectroscopy, J. Vac. Sci. Technol. B Microelectron. Nanom. Struct. 5 (1987) 530. https://doi.org/10.1116/1.583944.

[19] Common Bond Energies. http://www.wiredchemist.com/chemistry/data/bond_en.

[20] T. Kudo, M.S. Gordon, Theoretical studies of the mechanism for the synthesis of silsesquioxanes. 1. Hydrolysis and initial condensation, J. Am. Chem. Soc. 120 (1998) 11432–11438. https://doi.org/10.1021/ja980943k.

[21] Y. Du, X. Du, S.M. George, Mechanism of pyridine-catalyzed SiO_2 atomic layer deposition studied by fourier transform infrared spectroscopy, J. Phys. Chem. C. 111 (2007) 219–226. https://doi.org/10.1021/jp0638484.

[22] A.S. Ichimura, W. Lew, D.L. Allara, Tripod self-assembled monolayer on au(111) prepared by reaction of hydroxyl-terminated alkylthiols with $SiCl_4$, Langmuir. 24 (2008) 2487–2493. https://doi.org/10.1021/la703196b.

[23] P. Zemek, M. Corporation, PPM-Level HCl Measurements from Cement Kilns and Waste Incineratorsby FTIR Spectroscopy, 92626 (n.d.) 1–10. http://www.midac.com/files/AP-212.pdf.

[24] J.L. Doménech, B.J. Drouin, J. Cernicharo, V.J. Herrero, I. Tanarro, The high-resolution infrared spectrum of HCl+. The Astrophysical Journal Letters, 833 (2016) L32. https://doi.org/10.3847/2041-8213/833/2/l32.

[25] G. Lucovsky, S.Y. Lin, P.D. Richard, S.S. Chao, Y. Takagi, P. Pai, J.E. Keem, J.E. Tyler, Local atomic structure of silicon suboxides (SiOx, $x < 2$), J. Non. Cryst. Solids. 75 (1985) 429–434. https://doi.org/10.1016/0022-3093(85)90253-4.

[26] W.A. Pliskin, Comparison of Properties of Dielectric Films Deposited By Various Methods., J Vac Sci Technol. 14 (1977) 1064–1081. https://doi.org/10.1116/1.569413.

[27] M. Stadtmüeller, Mechanical Stress of CVD-Dielectrics, J. Electrochem. Soc. 139 (1992) 3669–3674. https://doi.org/10.1149/1.2069141.

[28] E. Issa, H. Nagel, J. Bartsch, M. Glatthaar, E. Rädlein, Application of hydrosilane-free atmospheric pressure chemical vapor deposition of SiO_x films in the manufacture of crystalline silicon solar cells, Thin Solid Films. 713 (2020). https://doi.org/10.1016/j.tsf.2020.138338.

6 Applications of the APCVD SiO_{2-x} films in PV fabrication

6.1 Introduction

The manufacture of silicon solar cells usually includes numerous processes that necessitate a high amount of chemicals and high energy consumption as well as vacuum tools. By way of examples: Si surface texturization, plated metallization associated with applying insulating layers to enable local plating, and plated metallization after local laser ablation of passivating layers.

Conventional c-Si texturization is, due to immersing completely in texturing solutions, executed on both Si wafer sides. Subsequently, polish etching of the wafer rear side is performed to remove most of the texture, which is often accompanied with the removal of unintended diffused emitters. Partial planarizing of the rear side is an essential step to enhance the open circuit voltage when mono-facial solar cells are to be fabricated. The wet polishing process is performed with a mixture of hydrofluoric acid (HF), nitric acid (HNO_3) along with either sulfuric acid (H_2SO_4) or potassium hydroxide (KOH) solutions, using inline polishing systems [1,2]. That results in etching of about 10-12 μm of the textured silicon surface. A faster process is the single side texturing performed in inline tools, thereby the wafers are mounted in a way that only their bottom face is in touch with an adjusted level of the etching solutions of HF, HNO_3 and acetic acid (CH_3COOH) solutions [3]. Other authors developed a dry technique for single side texturing with the so-called plasma-less atmospheric pressure dry texturing [4]. Another approach was investigated by coating the Si back side, either with SiO_2 shield grown by means of oxidizing its top surface or with SiN_x layers prepared by means of plasma enhance chemical vapor deposition (PECVD) technology [5–7].

Silicon heterojunction (SHJ) solar cells metallized by electroplating, as an alternative to screen printing, is currently of great interest [8,9]. For that matter, numerous processing sequences were investigated. For instance, by depositing a valve metal (characterized by self-passivating surface) like Al onto the transparent conduction oxide (TCO) layer of a SHJ solar cell. A grid pattern is then defined by printing a seed layer of e.g. Ag, which is plated later with Cu. Ultimately, the valve metal layer is etched off in a wet-chemical solution leaving the intended grid pattern and exposing the TCO layer in-between [10]. Another way is to screen-print a large area organic mask, excluding a grid pattern, on top of the TCO layer, followed by plating the open grid area and subsequently stripping the organic mask [11]. In order to save organic materials and avoid their disposal or recycling in large amount, another approach was developed to screen-print a grid pattern of organic materials in the shape of the final electrode [12]. The whole cell surface is then coated with a dielectric layer of e.g. PECVD SiO_x and eventually broken selectively in the patterned area which exposes the TCO surface to the plating electrolyte.

Kaneka cooperation developed a low silver consumption process for electroplating grid electrodes on top of SHJ solar cells [13,14]. Instead of applying an organic grid they replaced it by screen-printing a seed grid of Ag and then coat the cell surface with PECVD SiO_x plating mask, which is broken later selectively on top of the Ag seed, onto which a local plating is permitted. The advantage of this plating solution is that sputtering and subsequent etching of the sputtered valve metal used in ref. 10 is avoided. Moreover, the removal of the large-area organic mask and its related post-treatment are avoided, as well. Consequently, there is no need for using evacuated physical vapor deposition (PVD) tools nor to use dilute alkaline and acidic solutions. This approach, however, possesses a disadvantage of a considerable contact resistance between the TCO and finger lines because of the binders included in the low-temperature cure Ag paste that form the seed grid.

Passivated emitter and rear contact solar cells (PERC), recently raised as the dominant technology in the solar cell industry [8,15,16]. The standard technique to metallize such cells is screen-printing of an Ag paste on top of the passivation layer and then annealing the cell in order the metals (at the top and the back sides) to spike through and create an ohmic contact with the silicon substrate. The passivation layer employed is basically made of PECVD SiN_x which is utilized as an anti-reflection coating (ARC) of the cell, as well. Numerous processing sequences were intended to metalize the PERC solar cells with electroplating techniques making use of the SiN_x layer as a plating mask. This, however, collided with a major obstacle preventing the electroplating technology from wide distribution, namely parasitic plating (i. e. unintended plating) in the grid-free area. The parasitic plating on PERC solar cells is caused by different factors over the course of fabrication steps. S. Braun *et al.* [17] reported three significant mechanisms of parasitic plating generation in the SiN_x:H coated grid-free regions:

i) Insufficient texturing and hence resulting mechanical stress induce remained saw damage and 3D cracks in the pyramid valleys, thereby emitters exposed to the wet plating solution.
ii) Impurities remaining on the post-textured wafer surface such as residual Si nano/micro-fragments, therefore they act, after the emitter diffusion, as seeds for later metal plating. This was proven by indicating incomplete coverage of the emitter region with the nitride layer, where these residues are in contact with the wet plating solution.
iii) Non-uniformly deposited nitride layers end up with about 200 nm uncoated spots of the emitter, explicitly on the pyramid sides. This is attributed to possible contamination on the emitter surface, by what defects such as interruptions in the deposited nitride layer are present.

Some processing schemes can be adopted to avoid parasitic plating on the SiN_x ARC passivation layer. For instance, by avoiding the primary causes mentioned in ref. 17 or by retaining the native SiO_2, resulting after opening the grid pattern, on both emitter and nitride defects. That is in order to eliminate the plating through the SiN_x pinholes without ruining the subsequent plating [18]. Some other approaches are explained in [19,20], taking into account that parasitic plating may occur as a consequence of handling issues e.g. during cell fabrication unintentional scratches on the ARC may appear resulting in randomly uncovered emitter, and thereby unintended plating in the grid-free region.

In this chapter we introduce the newly developed APCVD SiO_{2-x} film as a cost-effective option to overcome the above-mentioned solar cell fabrication-restrictions. A first application is to utilize the APCVD SiO_{2-x} film as a texturing barrier, by which the high deposition temperatures and vacuum tools are avoided, meanwhile, the anti-etchant effect of the layers mentioned in ref. 5-7 are maintained. In a second application, the APCVD SiO_{2-x} films are attempted to act as a plating mask for SHJ solar cells with the assistance of self-developed organic paste that serves as a grid pattern, similar to ref. 12. A third APCVD SiO_{2-x} film application is to act as a plating mask of SHJ solar cells with the assistance of Ag seed grid, similar to ref. 13, 14. A fourth application is to utilize the APCVD SiO_{2-x} film as a protection form parasitic plating on PERC solar cells. A fifth application is to eliminate parasitic plating onto standard bifacial solar cells by exploiting the APCVD SiO_{2-x} films as a dielectric shield.

Theses application experiments and results were published in the journal 'Thin Solid Films' in September 2020 [21] and at the European PV Solar Energy Conference and Exhibition held in September 2019 [22,23].

6.2 APCVD SiO_{2-x} masking film for single-side texturing of c-Si wafers

Four-inch *p*-type chemically polished and {100} oriented FZ c-Si wafers with a thickness of 200 μm were employed. SiO_{2-x} films were deposited on one side of the wafers with a thickness of 180 -190 nm. The other side of the wafers, that is intended to be textured, was left without coating. Since it was proven in chapter 5 that the APCVD SiO_{2-x} film resilience can be improved after calcination at 300° C for 1 min, the coated samples were subjected to annealing treatment on a hot plate in ambient air. After calcination the coated samples were immersed in the texturing solution. it was an alkaline aqueous solution prepared from KOH with a concentration of 1.5 wt% intermixed with 0.5 wt% of the organic additive Ultra 1228 provided by Innovative Chemie für Industrie und Umwelt in Berlin. This additive was essential for obtaining random pyramids during texturing. The aqueous solution was first heated at 80 °C on a hot plate and homogenized continuously with a magnetic stirrer driven by a rotating magnet inside the hot plate. The samples were subsequently processed in the solution for 6 min and, after texturing, were rinsed in deionized (DI) water to remove texture contamination, as possible. The process sequence is illustrated in Figure 6.1.

The textured samples were eventually characterized in order to assess the process as to shielding performance of the oxide film on the rear side and its stability against the texturing solution, as well as, texture quality and light trapping of the wafer front side. For that, the following tools were exploited: i) a Woollam M-2000 ellipsometer, ii) a Zeiss Auriga 60 scanning electron microscopy (SEM), iii) a confocal microscope of the kind 3D-Laser Microscope LEXT OLS4000 form Olympus, iv) and a hemispherical reflectance tool of the kind Lambda UV/Vis/NIR Spectrophotometer from Perkin Elmer, furnished with an integrating sphere. The as textured samples were measured at three different angles: 65, 70 and 75° under the ellipsometry. The wavelength inspected was in the range from 300 to 900 nm. As to SEM measurements, 5 kV acceleration voltage and 7.7 - 8 mm working distance were adjusted. The inspected wavelength of the hemispherical reflectance was in the range from 300 to 1000 nm.

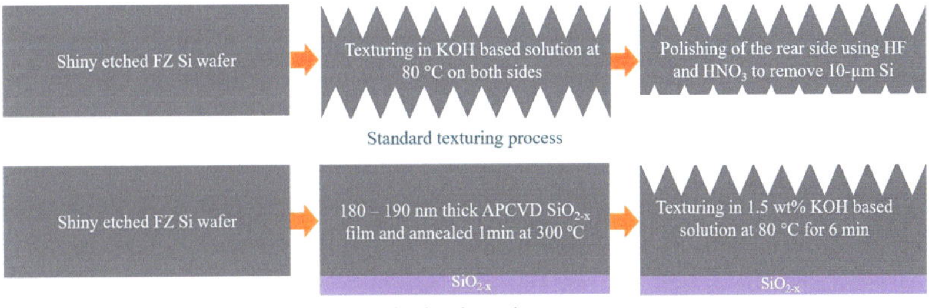

Figure 6.1: A pictorial representation of the conventional c-Si wafer texturing process in comparison with the newly developed process sequence utilizing APCVD SiO_{2-x} film applied on one side as protective shield from texturing etchants

SEM inspections shown in Figure 6.2 demonstrate that the rear side of the c-Si wafer (shown left) is completely protected from texturing etchants. No random pyramids are observed and the APCVD SiO_{2-x} coating is as smooth and compact as before texturing. On the other hand, the wafer's front side (shown right) is textured with the intended random pyramids featuring a height of 2-3 µm. No oxide residuals are observed between or on the pyramids.

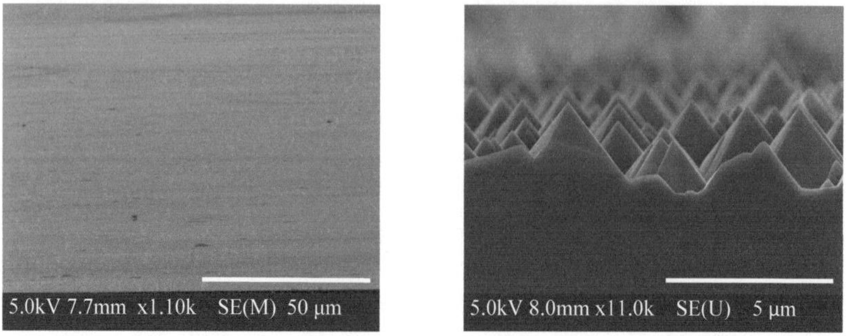

Figure 6.2: SEM images of a textured sample showing **(left):** a dense APCVD SiO_{2-x} protecting the rear side from texturing efficiently without any pyramids and **(right):** an unprotected front side of the same wafer with random pyramids and an absence of SiO_{2-x} residuals

Due to exposure to the texturing solution the oxide film was partially etched. The ellipsometry measurements performed on the oxide coated side of the wafer, before and after texturing, depicted that the annealed 180 nm thick APCVD SiO_x film was diminished to 80 nm after 6 min of texturing. 100 nm of the film was etched uniformly, yet the remaining 80 nm thick film was enough to protect Si rear side from texturing completely. Meanwhile, the uncoated wafer side was uniformly textured. Figure 6.3 left exhibits a macroscopic demonstration of the film protection resilience, as well as, the texturing uniformity, which are evident by the reflected and non-reflected picture on the sample upper sides, respectively.

Wrap-around of the SiO_x coating was inspected with a confocal microscopy all over the textured side of the Si wafers. As seen in Figure 6.3 (right), there is a narrow wrap-around of APCVD SiO_{2-x} on the outermost edge of the wafer textured side. However, most of the inspected surface is uniformly textured without APCVD SiO_{2-x} coating. The wrap-around was proven to be declined to 30 µm for all sample edges. This suggests a negligible influence on subsequent solar cell fabrication. Owing to the tight arrangement of sample holder-sample the deposition was shown to occur explicitly on the upper side of the wafer. Only a narrow wrap-around layer was present on the bottom side. In chapter 4 it was shown that this influence disappears when mounting a frame made of plastic around 6-inch wafers. However, since it was difficult to manufacture small frames suitable for 2 × 2 cm^2 wide samples or smaller, they were coated without using any protective frame.

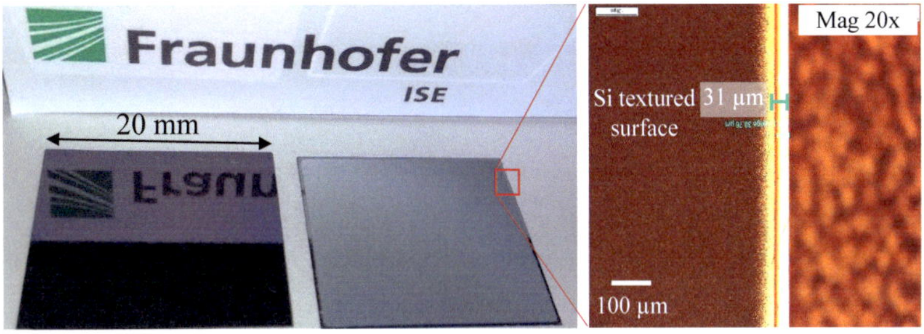

Figure 6.3: Left: Photos taken of the rear and front sides of a chemically polished FZ c-Si wafer after 6-min alkaline texturing at 80 °C showing a smooth SiO_{2-x} coated rear side (left) and uniform texturing on the front side (right). **Right:** An optical microscope image of a sample front side after texturing. A uniform texture is shown except for the outermost edge, which is coated with APCVD SiO_{2-x} suppressing the texture in a width of 31 µm.

An investigation of the performance of the single-side textured wafer was compared with a double-side textured FZ Si reference wafer. For that matter, a hemispherical reflectance was measured on both sides of a sample textured with the assistance of APCVD SiO_{2-x} coating. For comparison, the hemispherical measurement was performed on the double-side textured reference prepared by conventional texturing, as well. As to assessment of the APCVD SiO_{2-x} optical performance, it was compared with a simulated SiO_2 coated wafer back side. The simulation was performed using the transfer matrix method [24]. The simulation was performed assuming that the oxide film upon the c-Si wafer was 84 nm thick and its relevant refractive index and extinction coefficient were chosen the same as those of the thermal SiO_2 film [25].

Figure 6.4 depicts that in the wavelength range from 300 to 1000 nm the reflectance curve of the APCVD SiO_{2-x} on the rear side of the Si wafer was identical with the simulated one. When it comes to the textured side, it I shown that the reflectance curve of the single-side textured sample is in good match with that of the double-side textured reference. Consequently, the performance of the single-side textured wafer was proven to be similar to that of the conventional double-side textured wafers in the measured wavelength.

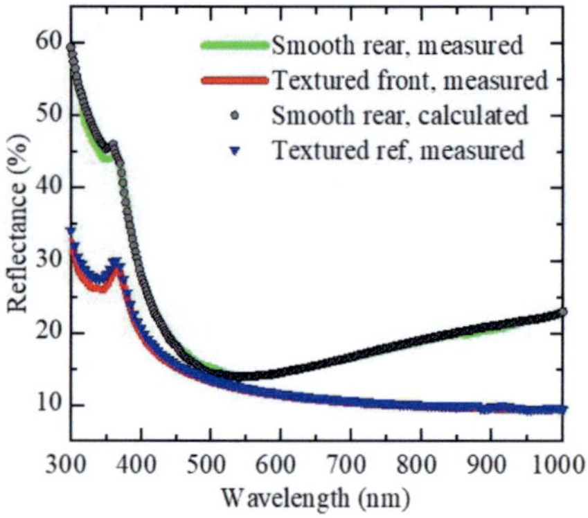

Figure 6.4: Hemispherical reflectance measurements of i) the front and rear side of an APCVD SiO_{2-x} coated FZ Si wafer and textured on one side ii) and the double-side textured reference FZ Si wafer. The calculated reflectance of a smooth Si wafer coated with 84 nm thick thermal SiO_2 film is also demonstrated.

An important step of this texturing approach was the removal of the deposited film from the Si surface, by which further processing of the rear side are possible. Etching of the film was, therefore, performed in 1 wt% HF solution. The measured etch rate of the SiO_{2-x} films after annealing at 300 °C for 1 min was measured to be 1 nm/s, thus etching of 84 nm thick film lasted for about 1.5 min.

In another investigation the annealing temperature of the APCVD SiO_{2-x} films was varied in accordance with the protection against texturing. About 90 nm thick oxide films were deposited on shiny etched FZ Si wafers of 2 × 2 cm^2 width after Radio Corporation of America cleaning. The deposition took place at room temperature and the $SiCl_4$-to-H_2O molar ratio

(MR) was about 0.52. The samples were, subsequently, annealed on a hot plate in ambient air, each for 9 min. The annealing temperatures were in the range from 160 to 320 °C with an interval of 20 °C. Ultimately, the samples underwent a texturing process at 80 ° C for 6 min following the same procedure mentioned above. Table 6.1 displays that all of the annealed films were unaffected by the texturing process. The sample annealed at 200 °C failed by accident as the samples annealed at temperatures lower and higher than 200 °C protected the Si surface sufficiently. The oxide coated surfaces remained smooth as before texturing. It was concluded that annealing the APCVD SiO_{2-x} films at relatively low temperature of 160 °C for 9 min was sufficient to protect from texturing etchants. Further investigation is recommended to check the lowest temperature and shortest annealing duration, at which the APCVD SiO_{2-x} film withstand the texturing process.

Table 6.1: Variation of the annealing temperature of the APCVD SiO_{2-x} films and their impact on the stability against texturing in KOH based solution for 6 min at 80 °C

	Sample 1	Sample 2	Sample 3	Sample 4	Sample 5	Sample 6	Sample 7	Sample 8	Sample 9
Temperature (°C)	160	180	200	220	240	260	280	300	320
Film thickness (nm)	90	90	90	90	90	90	90	90	90
Protected area (%)	100	100	60	100	100	100	100	100	100

6.3 APCVD SiO_{2-x} mask for plated metallization of SHJ solar cells

6.3.1 Plated metallization using a polymer grid pattern

The low investments of the metallization scheme shown in ref. 12 could be further reduced by means of proposing the newly developed APCVD SiO_{2-x} film as an alternative of the well-known high-cost PECVD SiO_x. As a proof of concept, a group of samples to be investigated were 2 × 2 cm^2 large alkaline textured SHJ solar cell precursors provided with a full-area Ag rear side metallization prepared by PVD. Figure 6.5 illustrates a schematic drawing of this process sequence as follows:

i) The experiments start with SHJ precursors of alkaline textured and 75 nm thick ITO coated c-Si wafers on both sides, optionally with or without metallized rear side.

Screen-printing of a polymer paste made of ethyl cellulose in the shape of the contact grid upon the front side of the cells. This is followed by drying the paste at 50 °C for 1 min on a hot plate in ambient atmosphere. The viscosity of the paste was fine tuned in order to permit sufficient masking of the ITO surface against the subsequent APCVD SiO_{2-x} film and to realize an optimally defined print image.

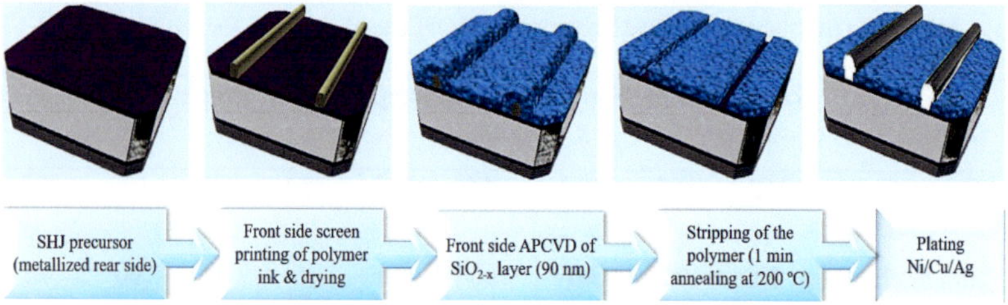

Figure 6.5: A schematic drawing of the process sequence for preparing of light-induced plated (LIP) SHJ solar cell front side metallization on ITO

ii) Depositing of about 90 – 100 nm thick APCVD SiO_{2-x} film all over the paste-patterned front side. This mask should feature an intact structure in the grid-free area, which is addressed by adjusting an MR of 0.5 – 1.

iii) Annealing the mask printed and oxide coated sample at ≤ 200 °C for 1 minute. This step is significant for evaporating the solvents from the paste inducing a shrinkage of the polymer finger and consequently breaking the oxide film on top, as shown in Figure 6.6. After the oxide film breaks sufficiently selectively in the grid area, stripping of the polymer paste is performed in an ultrasonic bath of action for 10 min at a temperature of 50 – 80 °C. Ultimately, the samples are rinsed in isopropanol followed by DI water so as to remove any paste residuals and SiO_{2-x} flakes.

iv) Light-induced plating [26] of a metal stack comprises Ni (at 50 °C), Cu and Ag (at room temperature) in the open area of the SiO_{2-x} plating mask. The Ni is deposited as a seed layer of 150 nm in thickness to promote the adherence of the Cu layer onto the ITO surface [27]. Subsequently, a 10 μm thick layer of Cu is electroplated for efficient electrical conductivity of the grid electrode. This is followed by plating a capping layer of 200 nm thick Ag for protecting the Cu from oxidation.

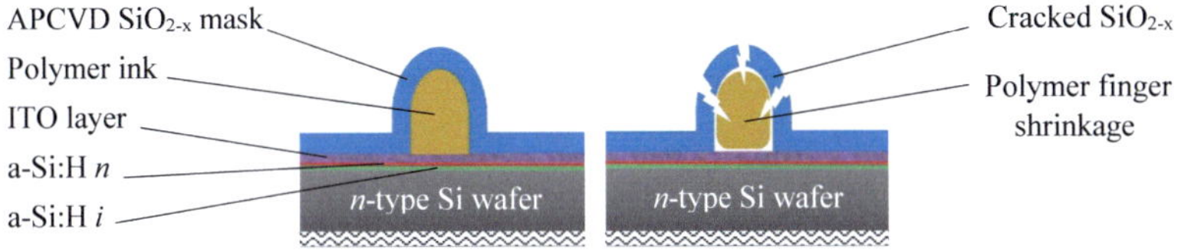

Figure 6.6: A pictorial cross-section of an ink finger screen-printed on the SHJ solar cell front side and coated with an APCVD SiO_{2-x} film. **Left:** Before annealing. **Right:** after annealing for 1 min at 200 C°. The back side of the cell is not represented in this drawing.

For characterization, a confocal scanning microscope of the LEXT OLS4000 3D-Laser kind was used to examine the printed polymer paste and the LIP of finger lines. An SEM of Zeiss Auriga 60 equipped with energy-dispersive X-ray spectroscopy (EDX) from Bruker Quantax 400 was used to inspect the oxide films during the process steps, as well as, the plated finger lines. EDX instrument was employed to analyze the paste residuals on the ITO surface after opening the finger lines. A Hewlett Packard flatbed scanner was exploited for the macroscopic scanning and assessment of the plated samples, as a whole. Eventually, in order to evaluate the adhesion of the plated metallization on ITO surface, a 3M tape test was conducted. The SEM acceleration voltage was 5 kV and the working distance was in the range 5 – 5.4 mm while the EDX acceleration voltage and working distance were 5 – 7 kV and 5.6 – 5.7 mm, respectively.

6.3.1.1 Investigation of the polymer mask

Different features of the polymer paste are required. One is a high aspect ratio, i. e. high and thin fingers. A finger width < 50 μm is preferred because this will form the openings for the subsequent plating process. A sufficient height of the fingers improves the subsequent stripping in solvents and ensures that the pyramid tips are not covered by the deposited APCVD SiO_{2-x}. The polymer fingers must, therefore, be higher than 3 μm to cover the pyramid tips and to provide enough bulk material to crack the APCVD SiO_{2-x} film deposited on top. The paste should feature a high thixotropic behavior in order to avoid fingers wider than the screen openings. Furthermore, the viscosity of the paste should be high enough in order to prevent spreading of the fingers once they are printed which leads to low and wide fingers, eventually. The reason for the spreading is the low surface tension of the paste on the ITO. This maximizes the contact area between the surfaces. An increase of the paste's surface energy leads to decreased spreading. On the other hand, the printed paste should be continuous without interruptions and therefore the paste with extremely high thixotropic behavior and viscosity is unfavorable. The most important requirement for the printed fingers is the possibility to be lifted off after coating with APCVD SiO_{2-x} films, which is a challenge. Recyclability of the paste is another goal.

Figure 6.7 shows optical microscope images of the ethyl cellulose-based polymer fingers screen-printed on alkaline textured and ITO-coated front side of SHJ solar cells. The polymer fingers feature different behaviors and paste properties:

a) A finger with spread edges and a width of 166 μm printed with a screen opening of 80 μm. The reason for that is the low thixotropic behavior of the utilized paste that does not recover after the removal of pressure applied by the screen. Moreover, the spread edges of the printed finger are attributed to relatively low viscosity.
b) A 1 – 2 μm high and 110 μm wide finger line insufficient for masking the textured and ITO coated surface against the subsequent APCVD SiO_{2-x} film as pyramid tips are uncoated. This finger is shown later to

induce difficult opening of the finger as a consequence of incomplete stripping of the oxide film on top.

c) A 1 – 2 μm high finger with a width of 135 μm because of extensive spreading of the paste on the ITO surface. The reason for the observation in Figure 6.7b, c, is the use of a paste with a too low viscosity.

d) A polymer finger line screen-printed on ITO coated surface with interruptions all along. This is due to a paste with too high thixotropic behavior.

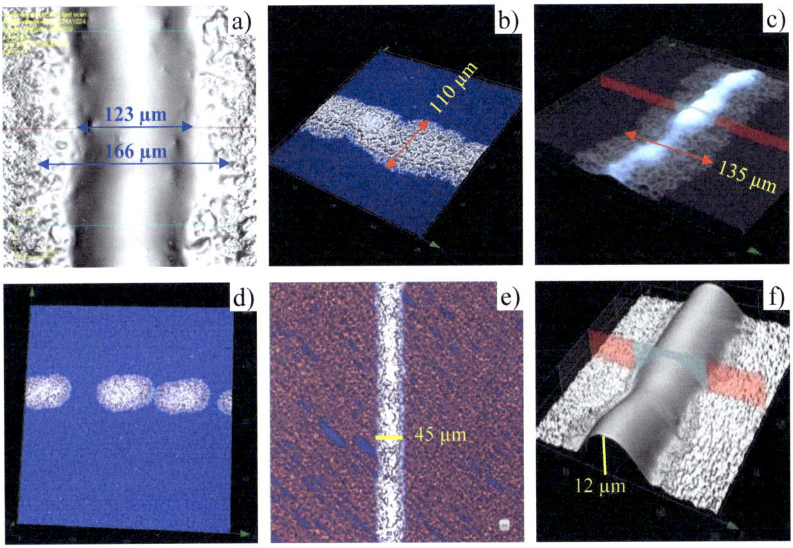

Figure 6.7: Confocal microscope images showing the topography of screen-printed ethyl cellulose-based polymer fingers on the front side of SHJ solar cell precursors. **a)** Low-thixotropic-behavior paste resulting in 166 μm wide finger, **b)** A polymer finger lower than 2 μm, **c)** A low-viscosity paste resulting in subsequent spreading over a wide area of 135 μm, **d)** An extremely high-viscosity paste developing finger interruptions, **e)** A fine polymer finger of 45 μm in width suitable for masking the ITO surface against SiO_{2-x} film, and **f)** 3D image representing the adequate height of the finger line in image e).

e) A finger line screen-printed from a paste with properly high viscosity and thixotropic behavior, delivered a finger width of 45 μm, only 5 μm wider than the opening of the screen.

A three dimensional (3D) optical microscopic image taken of the finger in Figure 6.7, e demonstrating a 12 μm high finger enough for protecting the ITO surface from the oxide layer. Low spreading and no finger disruptions were found. Even after several days of storing printed sample under atmospheric conditions, the finger profile had not changed, which is attributed to the paste with high thixotropic behavior and viscosity.

6.3.1.2 Microstructure of the APCVD SiO_{2-x} film

An inspection of the APCVD SiO_{2-x} film in the grid-free area was performed to check its conformity and integrity after annealing the polymer mask-coated SHJ precursors. The oxide plating mask to protect from electroplating should fulfill numerous characteristics. The masks were investigated in terms of i) appropriate thickness, ii) dense and compact structure, iii) intimate adherence to the ITO coated textured surface, iv) stability against annealing and alkaline/acidic solutions, v) and sufficient breakthrough voltage, known to characterize the SiO_2 films. The variation of APCVD SiO_{2-x} breakthrough voltage is not systematically investigated in this research. However, its efficient values are indicated by the stability of the films against subsequent electroplating voltage.

It was proven in chapter 5 that the SiO_{2-x} deposition rate (DR) (associated with the $SiCl_4$-to-H_2O MR) possesses a significant impact on the film quality. An important film quality is to feature high electrical properties. The microstructure of the SiO_{2-x} film reflects information about its electrical properties, e.g. when the film is porous and incorporates pinholes, its breakthrough voltage is assumed to be low, and thus a parasitic plating in the grid-free area is induced. The DR of SiO_{2-x} films applied on polymer grid-masked 2×2 cm^2 large SHJ samples was varied between 15 to 100 nm/min, and the film structure was subsequently inspected. The film thickness was also investigated in order to figure out the lowest thickness appropriate for protecting from parasitic plating. An SEM study was performed after annealing the oxide-coated samples at 200 °C for 1 min.

It was found that for SiO_{2-x} DRs > 70 nm/min, the film structure is granular with a rough surface. After annealing, the film was stripped from the ITO top surface, seen in Figure 6.8a. It was often observed that the rough APCVD SiO_{2-x} film features low adherence to rough surfaces such as ITO. This suggests that the SiO_{2-x} clusters, obtained at high deposition rates, are far larger than the ITO grain size and thus does not fill its superficial pores, see Figure 6.8b. Consequently, the SiO_{2-x} film adherence to the ITO surface is inadequate and, after annealing or during electroplating, it peels-off. This is found to cause parasitic plating later in section 6.3.1.4.

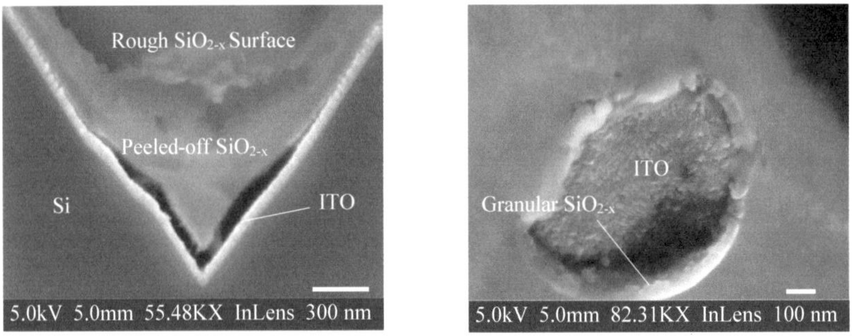

Figure 6.8: SEM images of alkaline textured and ITO coated SHJ solar cells front side showing **(left)** a cross-sectional view of an inferiorly thick APCVD SiO_{2-x} coating prepared at DR > 70 nm/min and peeled off at the pyramid velly after annealig at 200 °C, and **(right)** a Si pyramid coated with rough ITO and, on top, a agranular APCVD SiO_{2-x} film peeled-off after annealing at 200 °C

In the SEM study, where the film thickness was investigated, APCVD SiO_{2-x} films thinner than 70 nm revealed pinholes with 10 – 40 nm in width, see Figure 6.9 left. These pinholes permit the electrolyte to spike through the SiO_{2-x} and rich the ITO surface. Furthermore, it was found that when the SiO_{2-x} film is too thin and non-uniform, the pyramid tips are not protected enough from the electrolyte and susceptible to a high plating voltage. This hints that the smallest thickness forming a protective APCVD SiO_{2-x} mask from plating electrolytes should be larger than 70 nm.

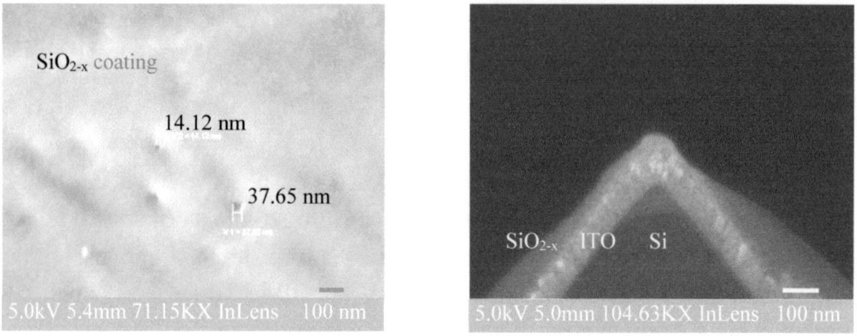

Figure 6.9: SEM images of alkaline textured and ITO coated SHJ solar cell front side exhibiting **(left)** a pyramid side coated with ≤ 70 nm thick SiO_{2-x} film containing through-pinholes, and **(right)** a cross section of a pyramid coated with SiO_{2-x} film non-uniformly at the sides and too thin at the tip

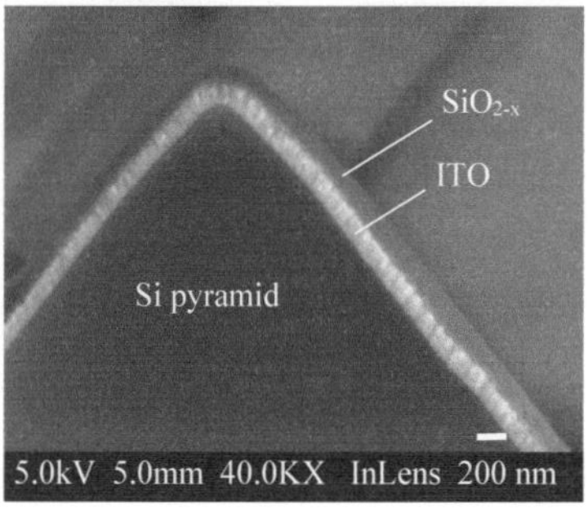

Figure 6.10: SEM image of alkaline textured and ITO coated SHJ solar cell front side showing a cross-sectional view of a dense APCVD SiO_{2-x} film prepared at DR ≤ 30 nm/min and remained intact and intimate with the underlying ITO layer after annealing at 200 °C

Figure 6.10 shows an SEM image of a cross section of an about 90 nm thick SiO_{2-x} film deposited on top of an alkaline textured and ITO coated SHJ solar cell front side. The film was deposited at a DR of 50 nm/min at room temperature. It can be seen that the APCVD SiO_{2-x} film is smooth and dense without pores, pinholes or cracks as opposed to the underlaying ITO layer which is relatively rough and granular. The SiO_{2-x} film was shown to coat the rough ITO layer uniformly and conformally without gaps and to remain resilient before and after annealing. The pyramid tips are well coated with the SiO_{2-x} film and thus the ITO layer underneath is protected from the alkaline plating solutions.

6.3.1.3 Structure of the APCVD SiO_{2-x} mask after polymer mask stripping

In this section stripping of the finger lines was investigated and the APCVD SiO_{2-x} mask structure was assessed, accordingly. The samples shown in Figure 6.7 were coated with APCVD SiO_{2-x} films and annealed at 200 °C for 1 min. Subsequently, the samples were subjected to a stripping process in an acetone path as explained previously. The developed samples were inspected under the laser confocal microscope LEXT OLS4000, as well as, with a Lumix digital camera from Panasonic in order to check the quality of the open finger lines. Figure 6.11a reveals that the polymer paste stripping is insufficient and the SiO_{2-x} films partially coat the pyramids. This is attributed to screen-printing of polymer finger lines (1 – 2 µm) lower than the pyramid heights prior to the oxide deposition. As a consequence, stripping of the polymer paste is difficult and the finger lines are incompletely open.

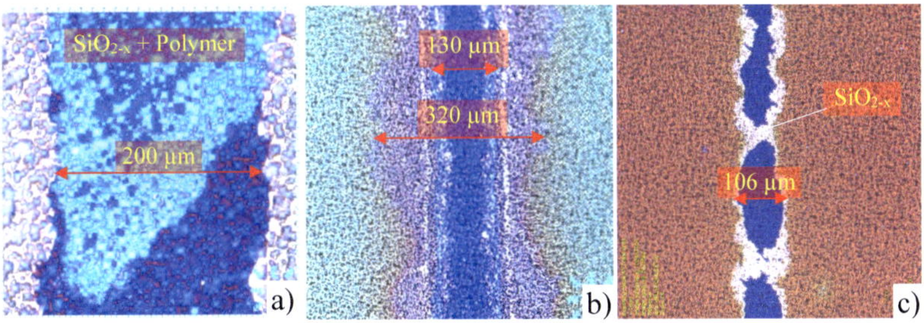

Figure 6.11: Laser confocal microscopic images of deficiently opened finger lines resulted from **a)** a screen-printed polymer finger lower than the pyramids' height, **b)** a cracked finger edges due to spreading of low-viscosity paste, and **c)** interruptions in the screen-printed polymer finger due to its high viscosity and thixotropic behavior

The residual polymer and oxide fragments are shown later to obstruct metal plating in the open fingers. Figure 6.11b reveals cracked finger edges, apart from the reasonable stripping in the middle. This consequence is attributed to the spreading of the polymer paste after screen-printing, which is due to the use of low paste viscosity. The spread paste at the screen-printed edges causes masking the pyramids partially. After deposition the SiO_{2-x} film coats the pyramid tips and intermixes with the thin paste at the finger edges. When the samples are annealed, the pyramids at the finger edge remain covered with a mixture of oxide flaks and polymer contamination. The finger edges are, therefore, not well opened similar to the case when screen-printing low polymer fingers. Figure 6.11c reveals interruptions in the finger line after stripping. This is ascribed to the original interruptions of the screen-printed polymer paste due to extremely high paste viscosity and thixotropic behavior. After deposition, the interruptions in the polymer fingers are coated with SiO_{2-x} layers and become difficult to lift-off.

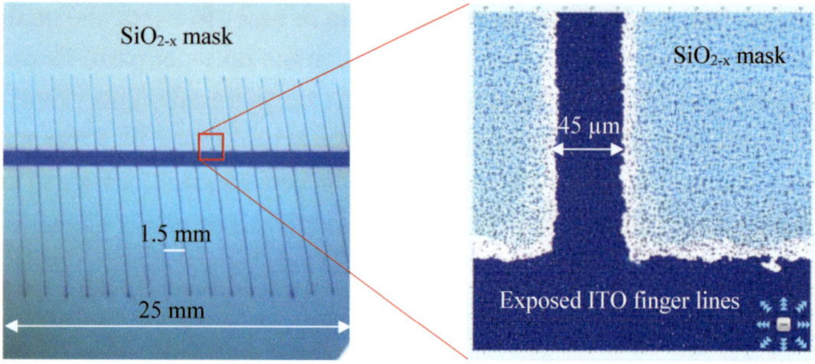

Figure 6.12: Left: A Photo of an APCVD SiO_{2-x} plating mask upon a SHJ solar cell front side after polymer lifting-off showing open fingers and bus bar. **Right:** a laser confocal microscopic image taken for the sample in (left) illustrating well defined and sufficiently open finger and bus bar

As for printed polymer finger-lines with sufficient height e.g. the 12 µm high fingers shown in Figure 6.7e, f, the SiO_{2-x} was completely impeded from reaching the ITO surface underneath. Annealing the samples facilitated cracking of the SiO_{2-x} film on top and thus enabled stripping of the polymer finger line adequately. A photo of the resulting SiO_{2-x} plating mask on front side of a SHJ solar cell is shown in Figure 6.12 (left). Owing to the eliminated spreading obtained with the sufficiently viscous polymer paste, the edges of the finger openings in the SiO_{2-x} mask were well defined, see Figure 6.12

(right). The width of the openings was nearly the same as that of the screen-printed polymer finger, i. e. about 45 μm.

SEM and EDX investigations (with an electron-beam diameter of 400 nm and a penetration depth of 300 nm) were performed in the open finger lines in the SiO_{2-x} mask to inspect the stripping quality and the ITO surface validation for subsequent metal plating. Figure 6.13 revealed the elements present in three open finger lines of three developed samples. The numbers in the EDX spectrum indicate the mass fraction of the elements detected. The inspected spots on the ITO surface are shown in the SEM images placed in the spectrum background. It was found that when the pyramids are insufficiently covered with the polymer mask, e.g. due to paste spreading, SiO_{2-x} fragments are deposited to the ITO surface, see figure 6.13a. When the polymer paste is inadequately lifted-off and the open finger lines are insufficiently cleaned with the solvent, polymer agglomerants remain on the ITO surface, see Figure 6.13b. In the case of well masking with the polymer paste and subsequently efficient stripping, the open finger was shown to be free of both oxide particles and polymer agglomerants. However, about 1.52 wt% carbon was detected in the EDX spectra, indicating that the paste was not completely removed, see Figure 6.13c.

It is worth mentioning that in the EDX spectra there were other elements indicated: In, O and Si, as well. This is attributed to the large penetration depth (300 nm) of the SEM electron beam and since the ITO thickness was about 75 nm, Si is also visible. The Sn content in the ITO layer was below the detection limit of EDX.

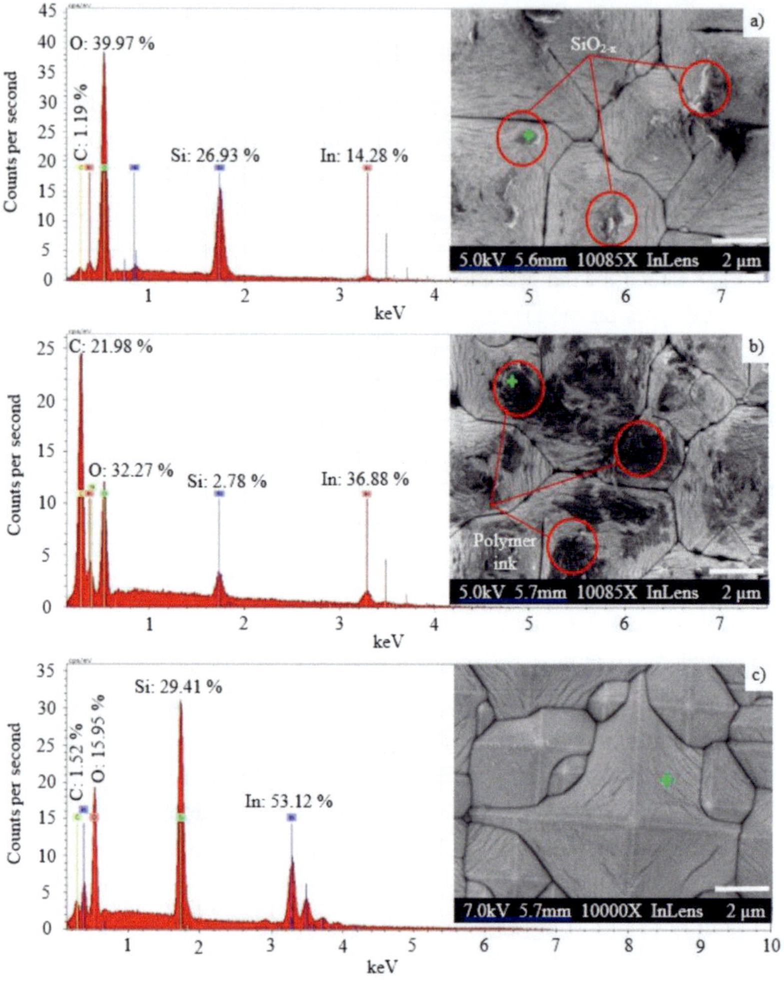

Figure 6.13: EDX spectrums and SEM images (in the background) of the ITO layer in the open area of the SiO_{2-x} mask after polymer stripping showing **a)** SiO_{2-x} fragments adhered to the ITO surface, **b)** polymer residuals adhered to the ITO surface, and c) ITO surface free of polymer agglomerants due to fairly stripped paste, with remaining negligible contamination. The Si presence in this spectrum is detected from below the ITO layer as the penetration depth of the electron beam is larger than the ITO thickness.

6.3.1.4 Performance of the SiO_{2-x} plating mask

It was found that when low-quality APCVD SiO_{2-x} mask is applied on the SHJ solar cell front side, no protection against the plating electrolyte (in this case alkaline solution) is provided. Consequently, not only does the plating take place in the open finger lines, but also in the grid-free area of the cell front side. Figure 6.14 exhibits an inefficiently plated front side of a SHJ solar cell full with parasitic plating. Part of the cell was not plated so as to clearly distinguish the parasitic plating of Ni. This influence can be attributed to many reasons, as previously mentioned, such as pores, peel off, and pinholes developed in the SiO_{2-x} films, besides the low resistivity against alkaline/acidic plating solutions.

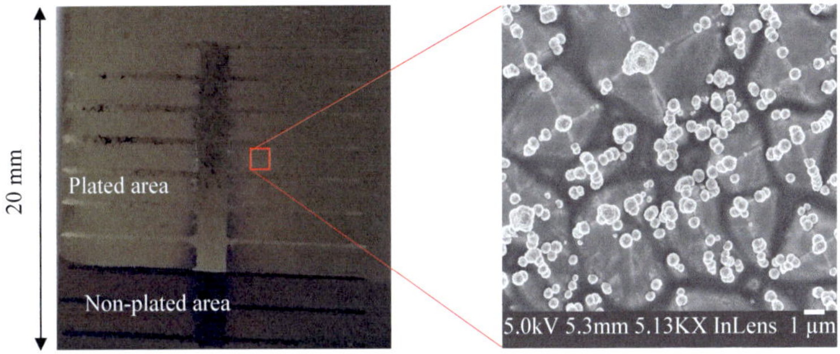

Figure 6.14: L eft: A photo of inefficiently plated SHJ solar cell front side showing severe parasitic plating with Ni in the grid-free area. **Right:** A SEM image taken for the plated cell shown left exhibiting parasitic Ni grains deposited on the pyramid tips, edges and sides, in an area coated an area coated with infirior SiO_{2-x}

It was also found that when extremely fine pinholes are present in thin films, a severe parasitic plating can take place in the grid-free area. A focused ion beam (FIB) was performed with the SEM tool to inspect the origin of metal grains grown on top of the SiO_{2-x} films. 40 slices were set vertically with a step of 50 nm to scan a Ni grain on a pyramid side. Figure 6.15 demonstrates a Ni grain grown on top of 50 nm thick SiO_{2-x} film coating the ITO layer. The Figure on the right is an enlarged area of the left figure where a metal root is observed. It was found that the parasitic Ni plating on SiO_{2-x}

films starts from a fine root of 10 – 40 nm in width and expand in 3 dimensional directions to form a large grain of a few micrometres. The origination of grain root is ascribed to fine pinholes in SiO_{2-x} film when they are thinner than 70 nm, as previously mentioned. The longer the plating process takes place, the larger the parasitic metal grains become, and hence the more the shading of the SHJ solar cell front side.

In the experiments where the quality of polymer finger stripping was investigated, the samples were subsequently metalized by means of LIP of Ni. In case of samples developed after applying a low-viscosity polymer mask, it was found that LIP does not take place all over the pyramid sides. Exhibited in Figure 6.16 left, one can see the pyramid tips free of LIP Ni. This is due to the coating of the tips with APCVD SiO_{2-x} layers, which is, in turn, because the polymer mask was thinner than the pyramids' height. Figure 6.16 right, reveals a simultaneous demonstration of incomplete Ni plating on the pyramids in the open finger line, as a consequence of contamination with both oxide fragments and residual polymer agglomerates. As a result, SiO_{2-x} flakes intermix with the Ni layer. The pyramid sides are not perfectly developed and this, in turn, causes interruptions and voids/impurities in the plated finger.

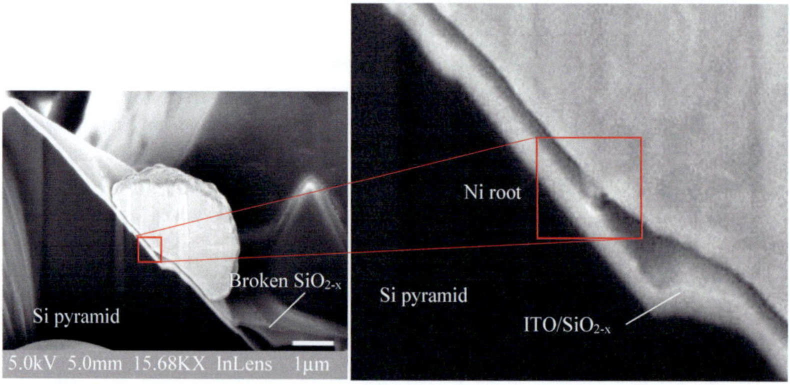

Figure 6.15: An SEM images of a Ni grain grown on a pyramid side coated with ITO and 50 nm thick SiO_{2-x} on top after FIB slicing. The image on the right is an enlarged slice area where the root of the Ni grain appears to stick out from the SiO_{2-x} thin film.

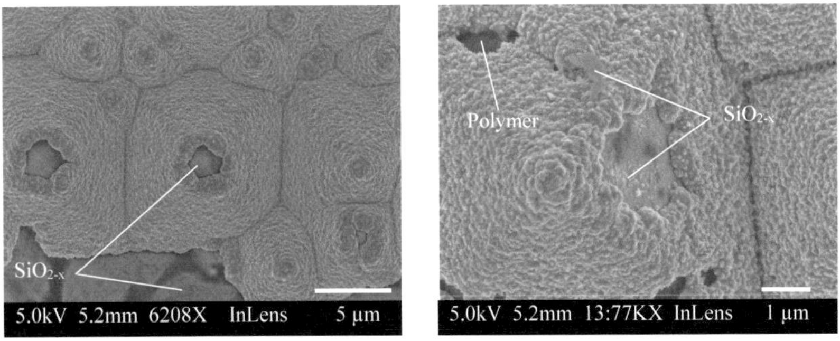

Figure 6.16: SEM images of LIP Ni on ITO coated pyramids after stripping the polymer mask in the regions where **(left)** the high pyramids are coated with the APCVD SiO_{2-x} layers at the tips and no LIP Ni is occurred, and **(right)** the polymer mask is incompletely stripped and the pyramid sides are contaminated with fall-off SiO_{2-x} fragments. Interruptions in the LIP Ni layer are shown.

This leads to a high contact resistivity of the metallized grid and, even worse, the adherence of the LIP Ni fingers diminishes because of the intermix with the SiO_{2-x} fragments imbedded on the ITO surface, as well as with the polymer residuals.

Other LIP experiments were performed for the samples where the polymer mask was stripped completely and no contaminants were present on the pyramids in the open finger lines. Theses samples were coated with about 90 nm thick SiO_{2-x} film completely covering the ITO in the non-grid area of the SHJ solar cell front side. The SiO_{2-x} films were shown to adhere conformally and durably on the rough ITO surface, as well as, to be dense enough for protecting the underlying ITO from unintended plating. Figure 6.17 shows a scan (left), using A Hewlett Packard flatbed scanner, and an SEM image (right) of a 90 nm thick APCVD SiO_{2-x} film on an alkaline textured and ITO-coated front side of a 2×2 cm^2 large SHJ solar cell after LIP of Ni, Cu and Ag. It is obviously shown that there is neither parasitic plating in the non-grid area nor interruptions in the plated finger lines.

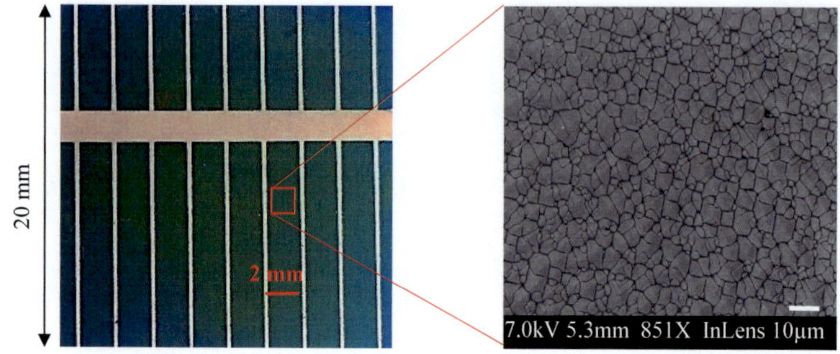

Figure 6.17: Left: Scan of the front side of a 2 × 2 cm² wide SHJ solar cell demonstrating an entire absence of parasitic plating in the grid-free area after applying local polymer and full-area SiO_{2-x} masks followed by 1 min annealing at 190 °C and electroplating of Ni, Cu and Ag stack metals. **Right:** SEM image taken for the cell shown left demonstrating an entire absence of metal grains on the pyramids coated with 90 nm dense SiO_{2-x}

The openings obtained after lifting off the screen-printed polymer fingers were about 40 – 45 μm wide, whereas the plated finger widths were larger in the range of 50 - 55 μm. This difference in finger width before and after plating is ascribed to the growth mechanism of the plated metal. In this plating approach, there are no barriers to restrict the metal growth to a direction perpendicular to the solar cell surface. Instead, a growth in 3-dimensional directions is induced. The SiO_{2-x} coated pyramids shown in the right image are completely clear without deposited metal grains or remarks of broken oxide mask.

In an additional experiment, a 3M tape test was performed to examine the adherence of the LIP finger lines to the ITO layer. Thanks to the thin anchor layer of Ni, the stacked-metal fingers did not peel off. This indicates that the small quantity of residual contaminants as detected by EDX causes a negligible effect on the adherence and structure of the plated metallization in the locally exposed ITO layer.

Illuminated I-V measurements were conducted to investigate the processed cells' performance. See Figure 6.18. In this proof-of-concept experiment the best efficiency obtained was about 12 % as accounted in

Table 6.2. However, reaching the state-of-the art efficiency depends on simultaneously numerous factors to consider such as the rheology properties of the used polymer ink, width of the screen-printed Ag fingers, edge recombination of the cut cell precursors, and functionality on 6-inch precursors. Apart from the APCVD SiO_{2-x} quality, the loss in V_{oc} shown in Table 6.2 is attributed to edge recombination since the cells were not isolated prior to the electroplating. The loss in I_{sc} is attributed to relatively wide and low height (about 1 µm) of the finger lines, due to which shading and high electric resistance contributed to the reduction of the processed SHJ cell efficiencies.

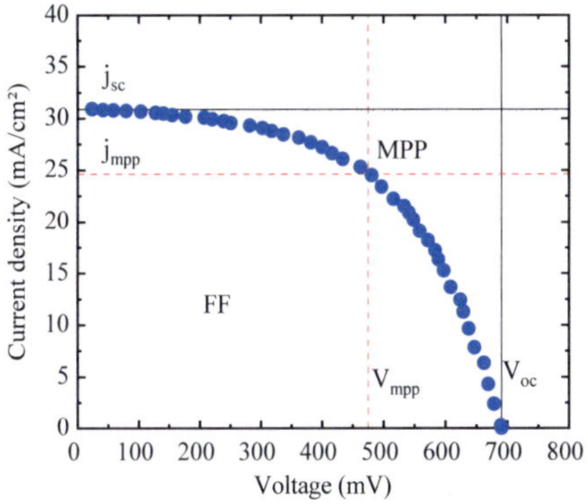

Figure 6.18: I-V curve of an electroplated front side of SHJ solar cell with an area of 2 × 2 cm² after APCVD SiO_{2-x} film coating and polymer mask stripping showing the voltage, current and MPP values

Table 6.2: I-V measurements of an electroplated SHJ solar cell with an area of 2 × 2 cm² after APCVD SiO_{2-x} film coating and polymer mask stripping in comparison with those of standard SHJ cells

Sample	Area	V_{mp}	J_{mp}	P_{mp}	V_{OC}	J_{SC}	FF	η
	[cm²]	[mV]	[mA/cm²]	[mW]	[mV]	[mA/cm²]	-	[%]
Standard	1	555,4	135,7	75,38	655,1	146,1	0,787	22
SiO_{2-x} masked	4,41	474,5	24,63	51,54	690,4	30,91	0,548	11,69

6.3.2 SHJ plated metallization using an Ag initiation layer

Plated front side metallization of SHJ solar cells was conducted by means of applying an Ag initiation layer to promote the subsequent Cu plating, similar to ref. 13, 14, and a full area APCVD SiO_{2-x} plating mask on top. A set of alkaline textured and ITO coated SHJ solar cell precursors with an area of 2×2 cm² featuring a full-area Ag rear side metallization obtained by PVD technique were employed. Figure 6.19 exhibits a schematic drawing of this process sequence, which is as follows:

i) Starting with precursors of alkaline textured and 75 nm thick ITO coated on both sides of c-Si wafers, optionally with metallized rear side or without (the rear side metallization is not depicted in this drawing).

ii) Screen-printing of a grid seed-layer using low cure-temperature Ag paste, as such subsequent Cu grid electrode is patterned with taking in to account that the width of the final plated finger lines will be larger than that of the seed one. Thus far obtained screen-printed Ag seed layer is as narrow as 30 – 40 µm. Therefore, high aspect ratios of the plated fingers are predicted. Besides considering their width, the screen-printed Ag fingers must as thin as 1– 2 µm, so that the material expenses are further reduced with maintaining the adherence of subsequently plated Cu to the ITO. After screen-printing of the Ag seed, curing the print is performed at 200 °C for 10 min, e.g. on a hot plate in the air. In terms of narrow finger seeds, other printing technologies were adopted to seed-printing of narrower Ag fingers such as inkjet printing. In [28], for instance, seed fingers featuring width of less than 20 µm are effectively inkjet-printed.

iii) Depositing of about 90 nm thick of an APCVD SiO_{2-x} film all over the Ag seed-coated ITO surface of the SHJ solar cell front side. $SiCl_4$-to-H_2O MR was in the range from 0.6 to 0.7. The determination of the oxide mask thickness is critical as it affects cracking of the APCVD SiO_{2-x} on top of the Ag seed. Two contradicted features should be acquired for the oxide mask: easy to break on top of the Ag seed and intact in the grid-free area. It is shown later in this section that the film thickness is the key factor to promote selectively thorough breaking of the oxide film.

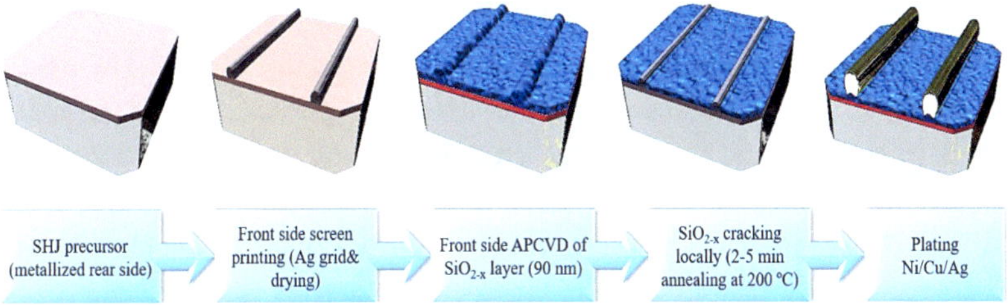

Figure 6.19: A schematic drawing of the process sequence for LIP metallization on SHJ solar cell front side after screen-printing of an Ag seed layer as a grid pattern. The rear side PVD Ag metallization is not represented in this drawing.

iv) Annealing of the sample at temperature less than 200 °C on a hotplate for 2 – 5 min. This step is essential as it delivers three functions. First, it causes cracking of the APCVD SiO_{2-x} film selectively on top of the previously printed Ag seed, thereby promoting subsequent Cu plating as the top Ag fingers is no longer insulated with oxide film. This impact is due to a significant change in the mechanical properties of the printed Ag fingers, namely the tensile stress built up. Consequently, at a defined thickness, the deposited oxide film does not stand this change and break down on top of the Ag surface. Second, annealing the masked samples enhances the APCVD SiO_{2-x} film stability against the plating voltage and the harsh alkaline/acidic plating solutions, as proven in chapter 5. Third, it further increases the adherence of the dry Ag paste to the relatively rough ITO surface, therefore it prevents the peel-off of the ultimately plated finger lines. After sufficient oxide cracking on top of the Ag seed it is preferred to rinse the sample in isopropanol followed by DI-water, so that oxide residuals are eliminated in the grid area.

v) LIP of Cu layer at room temperature on top of the exposed Ag seed grid. Because of the thoroughly exposed Ag fingers, in combination with adequately insulated ITO surface in the non-grid area, Cu plating takes place locally on the desired finger lines. The plated Cu grid electrode is subsequently coated with about 200 nm thin layer of Ag to protect the Cu surface from oxidation. In the course of plating process, the plating voltage was adjusted about 5 V, while the current density flowing through the cell was 20 mA/cm^2.

The samples prepared this way were subsequently inspected by means of SEM, EDX and macroscopically to check the performance of the oxide mask in both the grid-free area and on top of the Ag fingers after annealing, as well as, the ultimate finger structure after electroplating. SEM of Zeiss Auriga 60 equipped with EDX spectroscopy from Bruker Quantax 400 was employed. EDX instrument was exploited to analyze the elements in the plated fingers. A confocal scanning microscope of LEXT OLS4000 3D-Laser was used to check possible parasitic plating in the oxide mask. A Hewlett Packard flatbed scanner was used for evaluating the plated samples macroscopically. The SEM acceleration voltage was 5 kV and the working distance was in the range from 4.8 to 5.4 mm. The EDX acceleration voltage and working distance were 5 kV and 5.4 mm, respectively.

The APCVD SiO_{2-x} film thickness was varied as to its breakability on top of the cured Ag seed layer after annealing at 200 °C for 5 min, besides its resistance of the LIP process in the grid-free area. Figure 6.20 shows SEM images of two different finger lines of screen-printed Ag seed layers on top of the ITO after SiO_{2-x} film deposition and annealing. When the oxide film thickness is $\geq$ 250 nm (left image), it is insufficiently cracked and therefore the plated Cu contact includes interruptions since unbroken parts of the oxide isolate the Ag surface. On the other hand, when the oxide film thickness is $\leq$ 75 nm (right image), the oxide film is completely broken on top of the Ag surface and thus continuous plating of the Cu contact is obtained. Therefore, it was concluded that the thicker the APCVD SiO_{2-x} plating mask, the more durable it is and the more difficult it is to break on the top of Ag seed layer. Consequently, the plated Cu layer on the Ag seed layer becomes inhomogeneous.

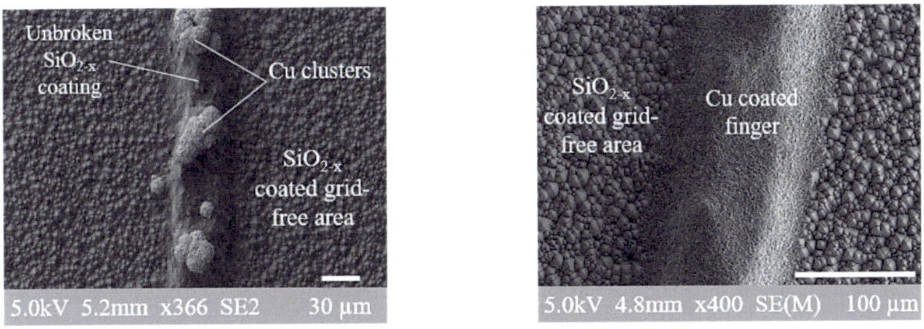

Figure 6.20: SEM images of scree-printed Ag seed lines coated with APCVD SiO_{2-x} films after annealing at 200 °C showing **(left)** inadequately cracked 250 nm thick SiO_{2-x} film on top inducing disruptions in the plated Cu contact, and **(right)** sufficiently cracked 75 nm thick SiO_{2-x} film on top leading to a continuous plating of Cu contacts

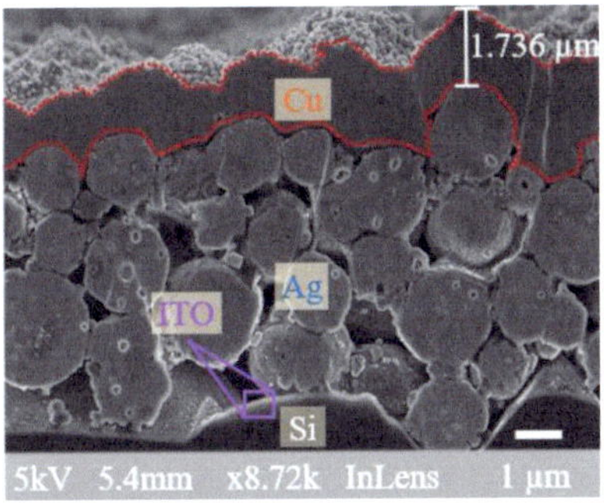

Figure 6.21: SEM image of cross-sectional view of a metallized finger line on the front side of a SHJ solar cell showing a screen-printed Ag layer coated uniformly with a dense LIP Cu layer. No SiO_{2-x} fragments are seen in the Cu/Ag interface nor in the Cu bulk.

In terms of plating quality and interface between the Ag seed and plated Cu, Figure 6.21 shows that after LIP process, a dense layer of Cu coating the Ag seed layer completely is obtained. No interruptions are found in the Cu layer proofing the absence of SiO_{2-x} residues that might remain after cracking the oxide film on top of the Ag seed. The Cu thickness measured throughout the finger cross-section is uniform, which is another evidence for the absence of SiO_{2-x} residues at the entire Cu/Ag interface. Furthermore, conformal contact between Cu and Ag layers is demonstrated. It is to be noted that this was a proof-of-concept experiment and therefore less care was taken of the exact thickness of metal layers as adjusted in ref 12, 13. In Figure 6.21 the LIP Cu layer is thinner and the screen-printed Ag layer is thicker than required.

In terms of APCVD SiO_{2-x} film durability and stability against LIP process, SEM and optical microscope characterizations were performed on SHJ samples after the plated metallization process-sequence mentioned above. Figure 6.22a exhibits a pyramid in the grid-free area coated with a 90 nm thick of oxide plating mask without metal grains or clusters. The APCVD SiO_{2-x} mask was shown to stand the electroplating voltage and to chemically resist the plating electrolyte, thus no cracks or peel-off occurred in the grid-free area. It was previously found that APCVD SiO_{2-x} films thinner than 70 nm do not protect the pyramids in the grid-free area from parasitic plating, when it comes to plating on SHJ solar cells after stripping the polymer mask. However, this thickness was less critical when it comes to plating on Ag seed layers. A threshold film thickness down to 55 nm was proven to protect the grid-free area from parasitic plating after annealing at 190 °C for 5 min. This is due to the fact that the Ag initiation layer increases the conductivity of the ITO in the oxide mask openings. Consequently, more electric current flows in the grid and less voltage, translated to an electric stress, is applied to the oxide film in the grid-free area. Annealing of the samples not only led to selective cracking on top of the Ag seed layer, but also increased the breakthrough voltage of the SiO_{2-x} film due to relative densification, explained in details in chapter 5. Ultimately, the optimum thickness of the APCVD SiO_{2-x} plating mask was determined to be 70 – 100 nm. The optical microscope image shown in Figure 6.22b, taken for the same sample inspected in a), demonstrate that the screen-printed Ag-finger is plated with Cu, completely without interruptions. In the area around the plated finger, which is coated with the coated with 90 nm thick APCVD-SiO_{2-x} mask, no Cu dots are found. Figure 6.22c demonstrates macroscopic examination of the same plated SHJ sample. It reveals uniform LIP Cu on the Ag seed layer throughout the whole solar cell area. Furthermore, an entire absence of parasitic plating is observed in the area between the finger lines. In addition to its role as a plating mask, the APCVD-SiO_x films are expected to play a protection role of plated SHJ cells from humid environments, as shown in Ref. 13, 14.

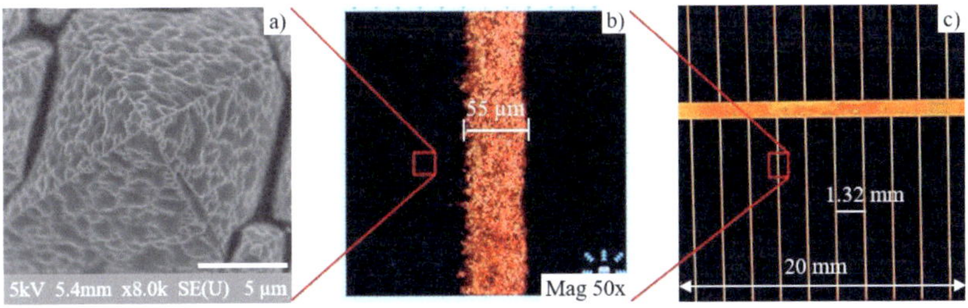

Figure 6.22: a) SEM image taken for the top view of a 90 nm dense APCVD SiO_{2-x} coated pyramids of a SHJ solar cell after annealing at 190 °C for 5 min and LIP of Cu, elucidating the absence of plated Cu clusters. The oxide mask is compact and adhere intimately on the rough ITO surface. **b)** Optical microscope image of a LIP finger line on the sample shown in a), revealing the absence of plated dots in the peripheral region of the Cu plated finger line. The left edge of the Cu plated finger line is irregular because of the overpressure applied from right to left when the Ag seed layer was screen-printed. **c)** A scan of the front side of the same SHJ solar cell front side seen in a), demonstrating the complete picture of LIP effect. An entire absence of Cu parasitic plating in the non-grid area is observed and no interruptions in the plated grid are seen.

6.3.3 Optical impact of the APCVD SiO_{2-x} films on the SHJ solar cell

In the aforementioned two approaches to the SHJ front-side plated metallization, no optical deficiencies are to be induced by maintaining the SiO_{2-x} film upon the TCO layer, irrespective of the module design. This is due to the fact that the refractive index of SiO_{2-x} (n = 1.41 -1.46) is lower than that of TCOs (n = 1.7 - 2) [29] in the relevant wavelength range from 280 to 1200 nm, so that more light intensity is trapped to the Si solar cell. Consequently, an increase in the short circuit current density of the solar cell is presumed [29,30]. On the other hand, if the solar cell is mounted in an encapsulant, the optical effect of SiO_{2-x} coating is covered up because its refractive index is nearly the same as the one of the encapsulant which usually is 0.5 mm thick ethylene vinyl acetate (n_{EVA} = 1.51) + glass plate.

Apart from depositing porous transparent oxides e.g. SiO_x onto the top surface of glass encapsulant [31–34], it is proposed to make further use of the newly developed APCVD SiO_{2-x} film such as an ARC onto the solar cells that will be ultimately encapsulated. Two methods are worth investigating: i) by decreasing the TCO thickness e.g. down to 20 nm (in the SHJ solar cell kind) [14] and ii) by providing the solar cell surface with multilayered ARC featuring a total refractive index lower than that of the solar cell top layer and higher than that of the encapsulation material (in all kinds of solar cells). The first proposal is quite possible since the ITO layer thickness can be decreased by etching in HCl solution.

The optimal refractive index and relevant thickness of the ARC can be calculated according to the following equations [24]:

As for a single-layer ARC:

$$n_1^2 = n_{si}.n_{air} \tag{6.1}$$

$$d_1 = \frac{\lambda_{air}}{4n_1} \tag{6.2}$$

As for a dual-layer ARC:

$$n_1^3 = n_{si}^2.n_{air} \tag{6.3}$$

$$n_2^3 = n_{si}.n_{air}^2 \tag{6.4}$$

$$d_1 = \frac{\lambda_{air}}{4n_1}, \quad d_2 = \frac{\lambda_{air}}{4n_2} \tag{6.5}$$

As for a triple-layer ARC

$$n_1^2 = n_{si}.n_2 \tag{6.6}$$

$$n_2^2 = n_1.n_3 = n_{si}.n_{air} \tag{6.7}$$

$$n_3^2 = n_2.n_{air} \tag{6.8}$$

$$d_1 = \frac{\lambda_{air}}{4n_1}, \; d_2 = \frac{\lambda_{air}}{4n_2}, \; d_3 = \frac{\lambda_{air}}{4n_3} \tag{6.9}$$

where

n_{si} is the refractive index of the silicon bulk.

n_1, n_2 and n_3 are the refractive indices of the first, second and third ARC layer, respectively.

d_1, d_2 and d_3 are the thicknesses of the first, second and third ARC layer, respectively.

n_{air} is the refractive index of air ($n_{air} = 1$).

λ_{air} is the wavelength of the surrounding air.

At a wavelength $\lambda_{air} = 632.8$ nm $n_{si} \approx 3.87$ and $n_{ITO} \approx 2$ [29]. If applied on SHJ solar cells, the layer stack is considered as follows: Si, ITO, ARC, and EVA on top. As a result, the ARC is supposed to feature a refractive index of 1.74 and a thickness of about 91 nm. This is impossible with silicon oxide films, thus showing that the APVD SiO_{2-x} film as single-layer ARC on top of encapsulated solar cells is not functional and other layers with higher refractive indices should be used.

Attempts have been made to apply porous APCVD SiO_x films featuring low refractive indices on glass encapsulants. Further investigations are still needed.

6.4 APCVD SiO_{2-x} as protective film against parasitic plating on PERC solar cells

The APCVD SiO_{2-x} film was employed to protect the electroplated metallization on the front side of 6-inch PERC solar cells from parasitic plating. It was shown in chapter 4 that the APCVD apparatus enables coating a substrate area of 6 inch, completely. The uniformity over 6-inch wide area was reasonable expect for the four sharp edges of square-shape substrates. However, this uncritical drawback is minimized when the solar cells used are of rounded edges. Figure 6.23 (top) illustrates the process flow of standard electroplated metallization of PERC and other SiN_x coated solar cells.

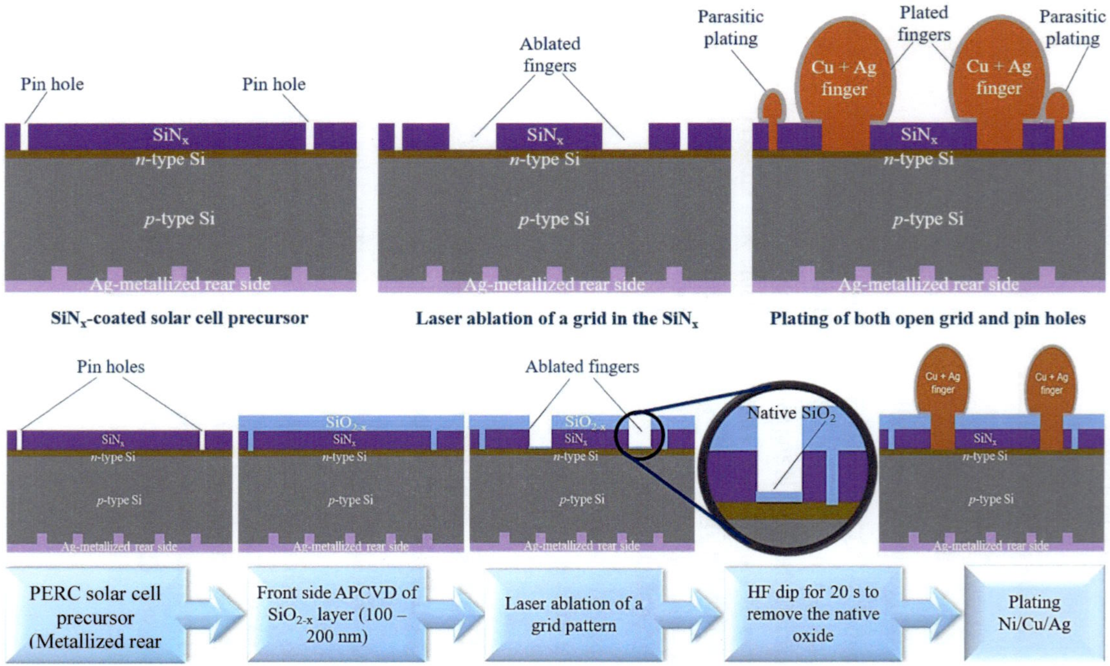

Figure 6.23: A pictorial representation of the process sequence for LIP metallization on the front side of a PERC solar cell after conventional local laser-ablation of SiN_x and HF dip for 20s (top) and after APCVD SiO_{2-x} film coating, annealing at 400 °C for 10 min, local laser-ablation of SiN_x, and HF dip for 20s (bottom)

In comparison with that, Figure 6.23 (bottom) illustrates the process flow of plated metallization on PERC solar cell front side by means of shielding the inferior SiN_x passivation/ARC layer from parasitic plating with the newly developed APCVD SiO_{2-x} film, preformed as follows:

i) Starting with precursors of alkaline-textured and phosphorous diffused *p*-type CZ Si solar cells featuring a SiN_x ARC on the front side and full area screen-printed rear side Al-metallization.

ii) Depositing of about 90 nm thick APCVD SiO_{2-x} film on top of the SiN_x ARC followed by annealing at 400 °C for 10 min on a hotplate in ambient atmosphere.

iii) Laser ablation of a contact grid pattern comprises 5 busbars and 110 fingers in order to expose the emitter on the Si surface locally.

iv) Etching of the native oxide formed after laser ablation in the treated locations by means of dipping the front side of the PERC solar cell precursors in 1 wt% HF solution for 20 sec.

v) Electroplating of the well-exposed grid with a Ni/Cu/Ag metal stack in inline plating tools by means of dipping the precursor front side in the electrolytes.

The quality of the emitter exposure in the obtained openings, and the shielding quality in the grid-free area were inspected under the optical microscope using a confocal LEXT OLS4000 3D-Laser kind. A Lumix digital camera from Panasonic was used to macroscopically assess both the thickness uniformity and protection quality of the annealed APCVD SiO_{2-x} films.

In order to demonstrate the difference between shielded and unshielded regions, the APCVD SiO_{2-x} film was deposited locally on the middle of a SiN_x coated PERC solar cell, as a proof of concept. Subsequently, laser opening, HF dip for 20s, and electroplating of the front side grid electrode were performed. Figure 6.24a shows a photo of one quarter of the PERC sample after electroplating. It can be seen that the APCVD SiO_{2-x} coated region is entirely free of parasitic plating in comparison with the uncoated region, wherein a massive parasitic plating took place. Looking at the handling scratches in the SiN_x layer on the coated and uncoated regions, it is seen that oxide shield efficiently covered the scratches and prevented parasitic plating, opposing the regions where no oxide is deposited.

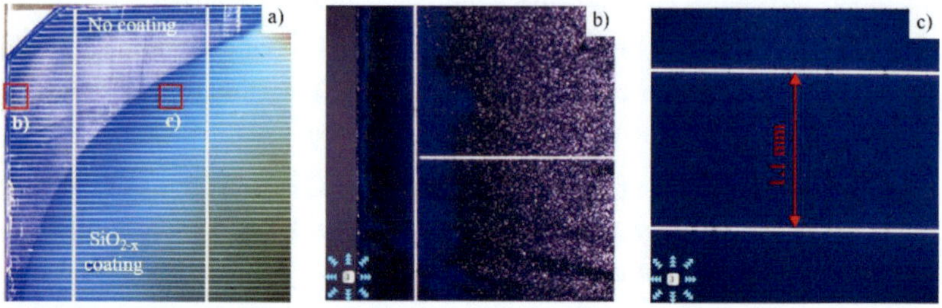

Figure 6.24: a) Photo of a PERC solar cell after electroplating of the front side Ni/Cu/Ag metallization. No parasitic plating is observed in the APCVD SiO_{2-x} coated region of the cell, whereas massive parasitic plating is seen in the outermost area where no APCVD SiO_{2-x} is deposited. b) Optical microscope picture taken for the uncoated area of the same solar cell shown in a) exhibiting massive metal dots SiOx. c) Optical microscope picture taken for the APCVD SiO_{2-x} coated area of the same cell revealing an entire absence of metal dots

For a closer look, Figure 6.24b and c, demonstrate optical microscope pictures of the uncoated and coated regions. The pyramids coated with the oxide shield were completely free of metal dots. An important observation to note is that the annealed APCVD SiO_{2-x} protective shield was only partially etched after HF treatment performed to remove the native SiO_2 layer that was generated after the SiN_x laser ablation to create the grid layout. This is due to the fact that post-deposition annealing enhances the APCVD SiO_{2-x} film stability against etchants, proven in chapter 5.

It is to note that the level of the PERC solar cell precursors was limited and therefore the performance of the processed cells was rather less than that reported in the state-of-the-art. The application explained above has been introduced as a proof of concept, for this research. However, the I-V measurements of the best PERC solar cell processed after front side full area APCVD SiO_{2-x} coating are shown in table 6.3.

Table 6.3: I-V measurements of an electroplated PERC solar cell after APCVD SiO_{2-x} film coating, laser ablation, and HF dip

J_{sc} [mA/cm^2]	pFF [%]	FF [%]	η [%]
39.38	80.55	77.83	19.12

6.5 SiO_{2-x} shielding from parasitic plating on bifacial standard solar cells

Irrespective of the substrate surface topography, another utilization of the new APCVD SiO_{2-x} was attempted anti-parasitic shield on the rear side of multi-crystalline bifacial solar cells. 6-inch large cells coated with SiN_x films containing primary scratches and metallized on the front side were tested. The same procedure performed for processing the PERC cells was repeated with changing in the oxide thickness and post-treatment. SiO_{2-x} films featuring thicknesses of 300 – 400 nm were uniformly deposited on the rear side and annealed at as low as 100 – 150 °C for 3 – 5 min. As shown in Figure 6.25, the SiO_{2-x} film prevented the parasitic plating on the SiN_x ARC as intended. As opposed to the cell unprotected with SiO_{2-x} film, the scratches were completely covered and the emitter was protected from getting in contact with the electrolytes. The optical microscope picture of the non-grid area, seen in Figure 6.25c, exhibits an entire absence of plated particles. In this investigation a trade-off between the oxide film thickness and both annealing temperature and time revealed that the thicker the APCVD SiO_2 film, the higher the protection against electroplating even at post-deposition annealing temperature lower than 150 °C. A possible explanation of this observation is that the breakthrough voltage of an insulator is proportional to the thickness, besides the $0.5 \leq SiCl_4\text{-to-}H_2O \text{ MR} \leq 1$ adjusted during APCVD SiO_{2-x} deposition. Subsequently, the newly developed APCVD SiO_{2-x} has been proven to effectively shield the SiN_x layers and eliminate the parasitic plating on the rear side of standard bifacial solar cells.

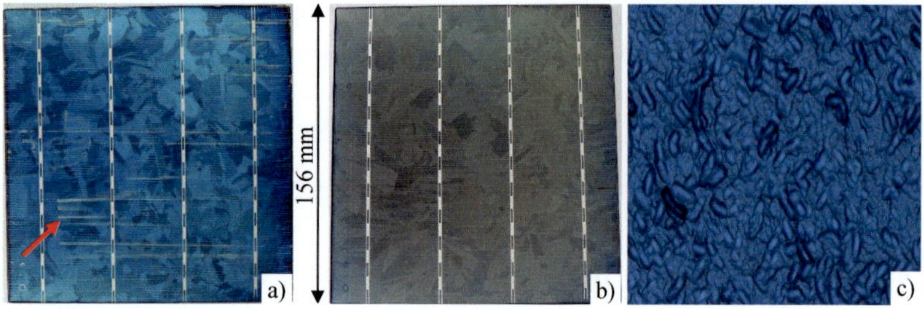

Figure 6.25: Photos of electroplated rear side of 6-inch iso-tropically textured and SiN_x coated bifacial multi-crystalline Si solar cells after **a)** laser opening and etching in 1wt% HF for 20s showing parasitic plating in the non-grid area and plated scratches in the SiN_x layer (the white horizontal lines are the plated scratches), and **b)** after SiO_{2-x} shield deposition, curing at 150 °C for 5 min, and HF dip for 20s showing that the SiN_x is completely coated with the APCVD SiO_{2-x} film and no parasitic plating is present in the non-grid area. **c):** An optical microscope picture of the same cell illustrating that no metal particles present on the SiO_{2-x} coated SiN_x surface shown in b), and no cracks in the non-grid area can be seen.

6.6 Summary

In this chapter, cost-effective process sequences for the manufacture of various kinds of solar cells were introduced. This was made possible by means of utilizing the inexpensive APCVD SiO_{2-x} films newly developed for various steps of PV fabrication. In one application APCVD SiO_{2-x} films were employed as protective coating against alkaline texturing solutions. A single-sided texturing of c-Si wafers was successfully obtained after coating the other wafer side with the newly developed APCVD SiO_{2-x} films. Annealing the oxide coated Si wafers at 300 °C for 1 min improved their durability and stability against the texturing solutions. The performance of the single side textured Si wafers was identical to that of the wafers textured with the conventional double side texturing technologies.

In another application the APCVD SiO_{2-x} films served as an electroplating mask in acidic electrolytes. A self-made polymer mask characterized by low cost investment was screen-printed in the shape of subsequent grid electrode on SHJ solar cells' front side and a 90 nm thick SiO_{2-x} film was deposited on top. The polymer and oxide masks were locally stripped after annealing at 200 °C for 1 min and ultrasonicating in solvents at 80 °C. After electroplating the open area on top of the SHJ solar cells, the oxide film was shown to sufficiently protect the area between the finger lines from parasitic plating. No interruptions in the plated finger lines were observed and the finger width could be minimized to 50 – 55 μm. The 3M tape test revealed a robust adherence of the plated metal onto the ITO top surface. Another scheme for SHJ plating metallization was investigated with the assistant of the APCVD SiO_{2-x} film after screen-printing of an Ag initiation layer in the form of the grid electrode so as to promote subsequent Cu plating. A 90 nm thick APCVD SiO_{2-x} film was selectively cracked on top of the Ag seed after annealing at 190 °C for 2 – 5 min permitting a complete coverage with LIP Cu. The Cu/Ag interface was free of SiO_{2-x} residuals allowing for uniform Cu layers plated conformally on the scree-printed Ag seed. Subsequently, the SiO_{2-x} mask applied on top, was shown to perfectly protect the SHJ solar cell front side from parasitic plating in the grid-free area.

In further applications, the APCVD SiO_{2-x} films were employed to prevent parasitic plating on the front side of 6-inch PERC solar cells. On top of the front side, SiN_x ARC characterized by structural defects was coated with about 90 nm of calcinated SiO_{2-x} film followed by laser ablation of a subsequent grid layout, HF dip for 20s, and LIP Ni/Cu/Ag metal stack. The SiN_x defects were adequately covered and no metal particles were detected on the gird-free region of the PERC solar cells after electroplating. In a similar processing sequence, structural defects and handling scratches present in the SiN_x ARC of 6-inch multi-crystalline bifacial solar cells' rear side were covered by the APCVD SiO_{2-x} film. 300 – 400 nm thick oxide films were deposited and annealed at 100 – 150 °C for 3 – 5 min. After Ni/Cu/Ag metal stack plating, the grid -free area and the scratches in the SiN_x layer were completely protected from parasitic plating.

In summary, it has been proven that the APCVD-SiO_{2-x} layer enables process sequences for various PV applications with low investment costs. No vacuum tools or plasma generators were used. Low Ag consumption and affordable metallization schemes for solar cell plating were successfully achieved and the shortcomings in solar cell plating metallization were finally overcome.

6.7 Bibliography

[1] M. Lippold, F. Buchholz, C. Gondek, F. Honeit, E. Wefringhaus, E. Kroke, Texturing of SiC-slurry and diamond wire sawn silicon wafers by HF-HNO_3-H_2SO_4 mixtures, Sol. Energy Mater. Sol. Cells. 127 (2014) 104–110. https://doi.org/10.1016/j.solmat.2014.04.006.

[2] R.P. J. Rentsch, R. Ackermann, K. Birmann, H. Furtwängler, J. Haunschild, G. Kästner, R. Neubauer, J. Nievendick, A. Oltersdorf, S. Rein, A Schütte, M. Zimmer, Wet Chemical Processing for c-Si Solar Cells - Status and Perspectives, in: 24th Eur. Photovolt. Sol. Energy Conf., Hamburg, Germany, 2009: pp. 1113–1117. https://doi.org/10.4229/24thEUPVSEC2009-2DO.1.1.

[3] G. Kästner, M. Zimmer, K. Birmann, F. Souren, J. Rentsch, R. Preu, Single side polish etching for the rear side of crystalline silicon wafers, in: 25th Eur. Photovolt. Sol. Energy Conf., Valencia, Spain, 2010: pp. 3–7. https://doi.org/10.4229/25thEUPVSEC2010-2CV.2.95.

[4] B. Kafle, A.I. Ridoy, P. Saint-Cast, L. Clochard, E. Duffy, K. Duncker, K. Petter, M. Hofmann, J. Rentsch, Atmospheric pressure dry texturing enabling 20% conversion efficiency on multicrystalline silicon PERC solar cells, in: AIP Conf. Proc., 2018. https://doi.org/10.1063/1.5049293.

[5] V. Prajapati, E. Cornagliotti, R. Russell, J.M. Fernandez, R.F. Clark, N. Stoddard, P. Choulat, J. John, High efficiency industrial silicon solar cells on silicon $MONO_2$ TM cast material using dielectric passivation and local BSF, in: 24th Eur. Photovolt. Sol. Energy Conf., 2009: pp. 1171–1174.

[6] S. Gatz, K. Bothe, J. Müller, T. Dullweber, R. Brendel, Analysis of local Al-doped back surface fields for high efficiency screen-printed solar cells, Energy Procedia. 8 (2011) 318–323. https://doi.org/10.1016/j.egypro.2011.06.143.

[7] P. Engelhart, N. Harder, R. Grischke, A. Merkle, R. Meyer, R. Brendel, Laser structuring for back junction silicon solar cells, Sol. Cells. (2008) 225–233. https://doi.org/doi.org/10.1002/pip.732.

[8] S.K. Chunduri, M. Schmela, High Efficiency Cell Technologies 2019 - From PERC to Passivated Contacts and HJT, TAIYANGNEWS. (2019).

[9] G. Beaucarne, G. Schubert, L. Tous, J. Hoornstra, Summary of the 8th Workshop on Metallization and Interconnection for Crystalline Silicon Solar Cells, in: AIP Conf. Proc., 2019. https://doi.org/10.1063/1.5125866.

[10] R. Rohit, A. Rodofili, G. Cimiotti, J. Bartsch, M. Glatthaar, Selective plating concept for silicon heterojunction solar cell metallization, Energy Procedia. 124 (2017) 901–906. https://doi.org/10.1016/j.egypro.2017.09.289.

[11] A. Khanna, K.U. Ritzau, M. Kamp, A. Filipovic, C. Schmiga, M. Glatthaar, A.G. Aberle, T. Mueller, Screen-printed masking of transparent conductive oxide layers for copper plating of silicon heterojunction cells, Appl. Surf. Sci. 349 (2015) 880–886. https://doi.org/10.1016/j.apsusc.2015.05.085.

[12] A. Aguilar, S.Y. Herasimenka, J. Karas, H. Jain, J. Lee, K. Munoz, L. Michaelson, T. Tyson, W.J. Dauksher, S. Bowden, Development of Cu plating for silicon heterojunction solar cells, in: 2017 IEEE 44th Photovolt. Spec. Conf. PVSC 2017, Portland, OR, USA, 2017: pp. 1–4. https://doi.org/10.1109/PVSC.2017.8366440.

[13] D. Adachi, T. Terashita, T. Uto, J.L. Hernández, K. Yamamoto, Effects of SiO_x barrier layer prepared by plasma-enhanced chemical vapor deposition on improvement of long-term reliability and production cost for Cu-plated amorphous Si/crystalline Si heterojunction solar cells, Sol. Energy Mater. Sol. Cells. 163 (2017) 204–209. https://doi.org/10.1016/j.solmat.2016.12.029.

[14] K. Yamamoto, K. Yoshikawa, H. Uzu, D. Adachi, High-efficiency heterojunction crystalline Si solar cells, in: Jpn. J. Appl. Phys., 2018. https://doi.org/10.7567/JJAP.57.08RB20.

[15] www.itrpv.net, International Technology Roadmap for Photovoltaic (ITRPV), Www.Itrpv.Net. (2019). http://www.itrpv.net/Reports/Downloads/2011/.

[16] A. Ciesla, R. Chen, S. Wang, J. Ji, Z. Shi, L. Mai, C. Chan, B. Hallam, C.M. Chong, S. Wenham, M. Green, High-voltage p-type PERC solar cells with anchored plating and hydrogenation, Prog. Photovoltaics Res. Appl. 26 (2018) 397–401. https://doi.org/10.1002/pip.2986.

[17] S. Braun, A. Zuschlag, B. Raabe, G. Hahn, The origin of background plating, Energy Procedia. 8 (2011) 565–570. https://doi.org/10.1016/j.egypro.2011.06.183.

[18] S. Kluska, A. Buchler, J. Bartsch, B. Grubel, A.A. Brand, S. Gutscher, G. Cimiotti, J. Nekarda, M. Glatthaar, Easy Plating-A Simple Approach to Suppress Parasitically Metallized Areas in Front Side Ni/Cu Plated Crystalline Si Solar Cells, IEEE J. Photovoltaics. 7 (2017) 1270–1277. https://doi.org/10.1109/JPHOTOV.2017.2720461.

[19] M. Heinrich, A. Kraft, M. Lieder, B. Hoex, A.G. Aberle, M. Glatthaar, Quantification of Front Side Metallization Area on Silicon Wafer Solar Cells for Background Plating Detection, Energy Procedia. 77 (2015) 717–724. https://doi.org/10.1016/j.egypro.2015.07.102.

[20] M.C. Raval, S.S. Saseendran, S. Suckow, S. Saravanan, C.S. Solanki, A. Kottantharayil, N_2O plasma treatment for minimization of background plating in silicon solar cells with Ni-Cu front side metallization, Sol. Energy Mater. Sol. Cells. 144 (2016) 671–677. https://doi.org/10.1016/j.solmat.2015.10.002.

[21] E. Issa, H. Nagel, J. Bartsch, M. Glatthaar, E. Rädlein, Application of hydrosilane-free atmospheric pressure chemical vapor deposition of SiO_x films in the manufacture of crystalline silicon solar cells, Thin Solid Films. 713 (2020). https://doi.org/10.1016/j.tsf.2020.138338.

[22] H. Nagel, E. Issa, T. Nagel, M. Glatthaar and S. W. Glunz, Hydrosilane-Free Low-Cost APCVD of SiO_2 Films for Crystalline Si Solar Cell Applications, in: 36th Eur. PV Sol. Energy Conf. Exhib., Marseille, France, 2019. https://doi.org/10.1007/0-387-30720-6_121.

[23] E. Issa, H. Nagel, L. Guzik, H. Javanbakht, E. Coron, E. Rädlein, M. Glatthaar and S. W. Glunz, Plated Front Side Metallization on Transparent Conducting Oxide, 36th Eur. Photovolt. Sol. Energy Conf. Exhib. (2019) 431–435.

[24] Max Born and Emil Wolf, Principles of Optics, Pergamon Press, Oxford, 1976.

[25] I.H. Malitson, Interspecimen Comparison of the Refractive Index of Fused Silica, J. Opt. Soc. Am. 55 (1965) 1205. https://doi.org/10.1364/josa.55.001205.

[26] Atteq ur Rehman and Soo Hong Lee, Review of the potential of the Ni/Cu plating technique for crystalline silicon solar cells, Materials. 7 (2014) 1318–1341. https://doi.org/10.3390/ma7021318.

[27] J. Geissbuhler, S. De Wolf, A. Faes, N. Badel, Q. Jeangros, A. Tomasi, L. Barraud, A. Descoeudres, M. Despeisse, C. Ballif, Silicon heterojunction solar cells with copper-plated grid electrodes: Status and comparison with silver thick-film techniques, IEEE J. Photovoltaics. 4 (2014) 1055–1062. https://doi.org/10.1109/JPHOTOV.2014.2321663.

[28] M. Glatthaar, R. Rohit, A. Rodofili, Y.J. Snow, J. Nekarda, J. Bartsch, Novel Plating Processes for Silicon Heterojunction Solar Cell Metallization Using a Structured Seed Layer, IEEE J. Photovoltaics. 7 (2017) 1569–1573. https://doi.org/10.1109/JPHOTOV.2017.2748999.

[29] S.Y. Lien, D.S. Wuu, W.C. Yeh, J.C. Liu, Tri-layer antireflection coatings (SiO_2/SiO_2-TiO_2/TiO_2) for silicon solar cells using a sol-gel technique, Sol. Energy Mater. Sol. Cells. 90 (2006) 2710–2719. https://doi.org/10.1016/j.solmat.2006.04.001.

[30] S. Zhang, Y. Yao, D. Hu, W. Lian, H. Qian, J. Jie, Q. Wei, Z. Ni, X. Zhang, L. Xie, Application of silicon oxide on high efficiency monocrystalline silicon PERC solar cells, Energies. 12 (2019) 1–11. https://doi.org/10.3390/en12061168.

[31] H. Nagel, A. Metz, R. Hezel, Porous SiO_2 films prepared by remote plasma-enhanced chemical vapour deposition - a novel antireflection coating technology for photovoltaic modules, Sol. Energy Mater. Sol. Cells. 65 (2001) 71–77. https://doi.org/10.1016/S0927-0248(00)00079-9.

[32] P.S. Nitride, Coatings for Planar Silicon Solar Cells using Remote, Sol. Cells. 260 (1999) 245–260.

[33] G. Helsch, E. Rädlein, G.H. Frischat, On the origin of the aging process of porous SiO_2 antireflection coatings, J. Non. Cryst. Solids. 265 (2000) 193–197. https://doi.org/10.1016/S0022-3093(99)00880-7.

[34] A. Inayat, B. Reinhardt, J. Herwig, C. Küster, H. Uhlig, S. Krenkel, E. Raedlein, D. Enke, Recent advances in the synthesis of hierarchically porous silica materials on the basis of porous glasses, New J. Chem. 40 (2016) 4095–4114. https://doi.org/10.1039/c5nj03591k.

7 Overall conclusions

In this research novel reactor design and method for the APCVD of SiO_{2-x} thin films utilizing $SiCl_4$ and H_2O vapors are investigated to serve for PV applications. The laboratory-scale design is characterized by low-investment APCVD tool as it is made of cheap materials such as polycarbonate and thermoplastic materials and no vacuum pumps, furnaces or plasma generators are used. Since it is manufactured with 3D printing technology, the reactor configuration can be tailored to suit variable substrate sizes and various applications. Moreover, it can be scaled-up to high-throughput solar cell production-lines. The deposition method is also cost-effective thanks to the tuned operation boundary conditions such as i) deposition at room or sub-room temperature, ii) long maintenance intervals, i.e. cleaning of the APCVD chamber is performed after 600 – 700 hours of laboratory operation, iii) low safety precautions since the $SiCl_4$ and HCl gases are not flammable and the exhaust HCl can be effortlessly neutralized with water, iv) and utilization of the cheap reactant precursors H_2O and $SiCl_4$. A simple calculation, made for the films deposited with the laboratory set up, reveals that the cost of reactant precursors is about 1 \$Cent per 100 nm thick SiO_{2-x} film on a 6×6 in^2 large Si wafer. Since this technique is contact free, e.g. no objects can get in contact with the precursors (apart from the deposition chuck), no sample breakage occurs as observed during laboratory operation. This technique is also metal/SiO_{2-x} particle contaminant free as the whole set up is made of plastic materials and no SiO_{2-x} dust is formed or adhere to the deposited films. Moreover, no corrosion or dissolvement takes place at the reactor parts or polymer-coated substrates as the $SiCl_4$ and HCl vapors do not react with such materials close to room temperature.

The motionless APCVD injector enables coating a 6×6 in^2 large substrate with the desired film except for insufficiently coated areas, located at the corners, forming 3 % of the total substrate surface. The SiO_{2-x} films are, however, deposited uniformly with a thickness variation value of ± 8 % all over the coated area. No wrap-around films or outermost thick rims of SiO_{2-x} are observed at the coated Si wafers. A wide range of deposition rates – from low ($\leq$ 1 nm/min) to high ($\geq$ 250 nm/min) depending on the film property – is possible thanks to the fine-tuning of the reactant molar ratio and the deposition temperature. In other words, the shortest time needed for depositing 100 nm (compatible for all investigated PV applications) thick SiO_{2-x} film upon 6 × 6 in^2 large Si wafer is about 24s. This makes the new technique a promising candidate for PV and other optic industries. No mechanical stress occurs in the as-deposited films nor the post-annealed films, and the films are deposited conformally on different surface structures, as depicted from the micro-scale inspection.

Post-deposition annealing is essential for treating the APCVD SiO_{2-x} films. Depending on the later application, the oxide film can be remedied by annealing at a temperature range 190 °C for 1 min – 400 °C for 10 min. To eliminate the hydroxyl impurities in the film, it is annealed at 600 °C for 15 min. However, this long-term impurity elimination is unnecessary for the demonstrated PV applications since it might prompt a tensile stress in the film deposited at molar ratios lower than 0.5 or higher than 1. On the other hand, annealing the films, deposited at molar ratios between 0.5 and 1, at 900 °C does not influence the film quality. Annealing at 400 °C decreases the APCVD SiO_{2-x} film refractive index slightly. However, after HF dip, the refractive index of the annealed film increases again slightly.

The newly developed APCVD SiO_{2-x} films are compatible for various PV applications. The films, with 180 nm in thickness, protect the Si wafer surface form alkaline texturing solutions, thus promote single side texturing when deposited on one side of the wafer. For this application it is enough to anneal the samples at 300 °C for 1 min or at 160 °C for 9 min. The oxide films act perfectly as an electroplating mask on the front side of SHJ solar cells. After coating with about 90 nm thick oxide, no parasitic plating is observed in the area between the finger lines. Annealing the samples before plating at 190 °C for 1 – 5 min, besides curing the oxide film, is sufficient to strip the printed polymer fingers and to crack the oxide layer on top of the Ag seed layers. The oxide films serve as a protective shield on the front side of PERC solar cells and the rear side of standard multi-crystalline bifacial solar cells against electroplated metallization. The parasitic plating is prevented and handling scratches are covered by 90 nm thick films annealed at 400 °C for

10 min for PERC solar cells, whereas for standard bifacial cells, the oxide films are 300 nm thick and the annealing is performed at 150 °C for 3 – 5 min.

8 Outlook

It is still an open question why the chemical stoichiometry of the APCVD SiO_{2-x} film is not perfectly O:Si = 2:1. More understanding of the reaction steps between the $SiCl_4$ and H_2O molecules is needed. The practical $SiCl_4$-to-H_2O molar ratio is investigated to be around 0.7, at which high deposition rate and film quality are obtained. This, nevertheless, contradicts the theoretical ratio (0.5) calculated form the hydrolysis equation of $SiCl_4$. It is recommended to investigate the quality of the gas mixture further, e.g. by means of increasing the mixing elements in the static mixer. Despite being reasonably uniform over a 97 % of 6 × 6 in^2 wide substrates, the current design of APCVD injector holds the potential to be improved further, by which substrate areas larger than 6 × 6 in^2 can be uniformly coated. The laboratory-type APCVD reactor allows for processing one sample per run. However, there is a high possibility to upscale it to the automated inline type. The current setup enables coating one side of the substrate. Experiments are recommended to obtain double-side coating per deposition run with a modified design of the APCVD injector. A promising proposal is to install a peripheral sample holder that supports the plane substrate at e.g. four outermost edges and permit enough distance from the injector's bottom base. This, consequently, allows for gas circulation on both substrate sides simultaneously. This is beneficial for affordably processing bifacial solar cells and various double-side-functioned microelectronic devices. Further investigation is suggested to ultimately assure the reasons behind the refractive indices 1.59 – 1.66 obtained at high molar ratios and deposition temperatures. A future investigation could be in the direction of tracing Ortho-silicic acid (featuring a refractive index of 1.535) in the as-deposited oxide films. Since FTIR measurements show an absence of Cl peaks, a highly sensitive tool is to be used to detect Cl bonds that might be triggered in the films causing the refractive index of 1.66. The etch rate experiments can be extended to investigate the uniformity of the APCVD SiO_{2-x} structure across the whole film thickness, before and after annealing. For this purpose, intermittent etching with a fixed stepwise can be performed and the film properties can be inspected after each etching step. It is worth researching an alternative chemical precursor to $SiCl_4$ for the deposition of APCVD SiO_{2-x} films. The alternative precursor should pose no APCVD operation risks and be as inexpensive as $SiCl_4$.

The APCVD SiO_{2-x} film possesses the potential for further applications in the future. Possible PV applications would be e.g. as i) capping for the Al_2O_3 passivation layer applied on the rear side of PERC solar cells, hence protection of the facing Si surface from alloying with the full area Al metallization during firing, ii) barrier for emitter diffusion, which could follow the single-side texturing step after maintaining the oxide film deposited on the rear side, iii) electroplated metallization mask for SHJ solar cells using different schemes from the introduced ones in this book, iv) and anti-reflection coating on SHJ solar cells, e.g. after mixing the SiO_{2-x} with TiO_2 to increase the film refractive index to the value 1.74 calculated in chapter 6. The TiO_2 material can also be deposited with the newly developed APCVD reactor. The mixed APCVD SiO_{2-x}/TiO_2 film can be used in optical fiber manufacturing as a material for the core. Because of its sufficient density, coverage conformity, thickness uniformity, and contaminant-free structure, the APCVD SiO_{2-x} film opens the doors broadly to be utilized in similar applications to PV ones, e.g. as a gate oxide in nano-microsystem technologies, protection for displays, etc. As it is possible to agglomerate porously, yet adhere durably and intimately to the substrate surface, the APCVD SiO_{2-x} could be a potential anchor interlayer to inspect upon various oxides, ceramics, etc. The dielectric strength of the SiO_{2-x} material as well as the corresponding breakthrough voltage is still to be investigated so as to allow utilization, e.g. in capacitors, wherein the SiO_{2-x} acts as the in-between insulator. It is proved through this book chapters that ultrathin SiO_{2-x} films of 1 – 2 nm are possible to obtain with the newly developed APCVD system. The films thinner than 70 nm are, nevertheless, characterized by considerable pinholes. If the ultrathin SiO_{2-x} films were improved to be deposited with negligible pinholes, they can be a competitive candidate to serve as a tunnel oxide in the so-called tunnel oxide passivated contacts (TOPCon) solar cells.

9 Appendix

9.1 XPS investigation of the APCVD SiO_{2-x} layers' stoichiometry

For comparison with the FTIR and EDX measurements performed in Section 5.2, in this chapter the Si:O atomic ratio in the same samples, which are APCVD SiO_{2-x} films on Ge wafers with and without annealing at 300 °C for 1 min, was investigated using X-ray photoelectron spectroscopy (XPS) [1]. With XPS, only the topmost 5 – 7 nm of the APCVD SiO_{2-x} layers are characterised because in solids the mean free path length of the photoelectrons generated by X-rays is small due to inelastic scattering. An ultra-high vacuum system from SPECS with a base pressure of 2×10^{-8} mbar was used for the measurements. This module is equipped with a hemispherical analyser from the company PHIOBOS. The high-voltage excitation energy (photo energy) was adjusted to hv = 1486.6 eV, the size of the measuring spot was around 1 mm and the system was first calibrated with a polycrystalline silver sample. The stoichiometric calculations were performed under the assumption of a homogeneous element distribution in the film volume using the so-called relative sensitivity factors. The samples were measured as loaded to the chamber and after sputtering with an argon ion beam for 20 min to remove possible C contaminants. The adjusted beam parameters were as follows: beam energy of 3 kV, sputtered area of 1 x 1 cm^2, and Ar pressure of 2.2×10^{-3} mbar. The measured binding energies were corrected by the C1s (C–C) peak assuming its main component is caused by the contaminant aliphatic carbon (H–C–H) at a binding energy of 284.8 eV. The calculation of Si:O atomic ratio was based on measured peak areas of the oxygen O1s (at 533 eV) and the silicon Si2p (at 104 eV) core-level state.

The measurements of as-loaded samples revealed Si and O peaks in the XPS spectra and minor C contaminants, see Figure 9.1. The Si:O atomic ratio, calculated before sputter cleaning, was 1:1.90 for both, the annealed and the non-annealed sample. This atomic ratio is in good agreement with the previously mentioned EDX measurements, where the Si:O ratios for the samples without and with annealing were 1:1.80 and 1:1.93, respectively. However, more investigations for annealed samples are recommended.

After sputtering, C was removed from the top surface of the film. The maximum of the Si peak shifted to the corrected value of the Si2p peak before sputtering. The peak asymmetry at lower binding energies increased for both samples, whereas the Si:O ratio decreased to 1:1.7 owing to the decrease in O concentration. A possible interpretation is that the detected OH concentration in the film is higher after sputtering. This might be an effect of possible preferential sputtering of O atoms or of cracking of Si-O bonds due to the bombardment with high-energy ions. That would explain the decrease of the O content as well as the stronger asymmetry of the Si2p peak at the same time. The presence of Si-OH bonds in the oxide film leads to photoelectrons indicating a binding energy lower than that of SiO_2, see Figure 9.1, but higher than that of Si-Si bonds [2].

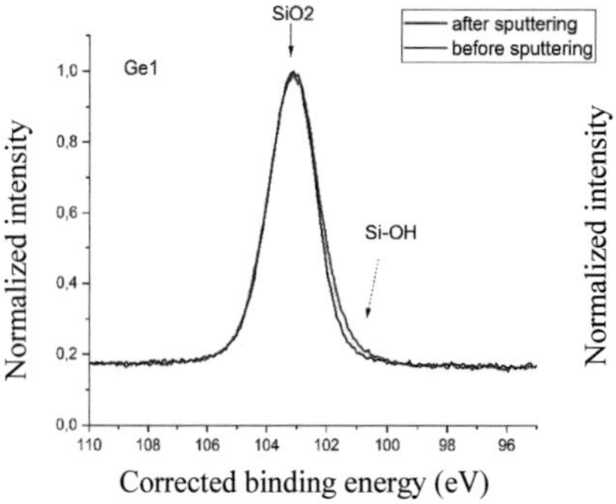

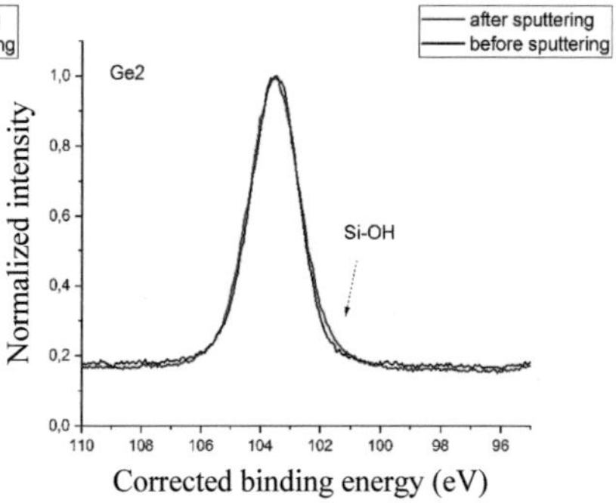

Figure 9.1: Measured high resolution XPS spectra of the topmost surface layer of silicon-oxide-coated Ge wafers, measured before and after sputtering. **Left**: Without annealing. **Right**: After annealing at 300 for 1 min. The spectra were normalized in order to better compare changes in the peak symmetry.

9.2 Bibliography

[1] S. Hüfner, Photoelectron Spectroscopy, Principles and Applications, Volume 82, Springer Series in Solid-State Sciences, Berlin, 1996.

[2] P. Post, L. Wurlitzer, W. Maus-Friedrichs and A. P. Weber, Characterization and applications of nanoparticles modified in-flight with silica or silica-organic coatings. Nanomaterials, 8 (2018), p. 530. doi:10.3390/nano8070530.

10 List of abbreviations and symbols

List of abbreviations

Abbreviation	Description
3D	Three dimensions
ABS	Acrylonitrile-butadiene-styrene
ALD	Atomic layer deposition
APCVD	Atmospheric pressure chemical vapor deposition
a-Si:H	Hydrogenated amorphous silicon
ATR-FTIR	Attenuated total reflection FTIR
BPSG	Borophosphosilicate glass
C-CVD	Combustion chemical vapor deposition
c-Si	Crystalline silicon
CVD	Chemical vapor deposition
DI	De-ionized
DRAM	Dynamic random-access memory
EDX	Energy dispersive X-ray spectroscopy
EVA	Ethylene vinyl acetate
FIB	Focused ion beam
FTIR	Fourier transform infrared spectroscopy
FTO	Fluorine-doped tin oxide
FWHM	Full width at half maximum
FZ	Float zone
FZO	Fluorine-doped zinc oxide
HF	Hydrofluoric acid
IR	Infrared
ITO	Tin-doped indium oxide
ITO:F	Tin:fluorine-doped indium oxide

Abbreviation	Description
LCD	Liquid crystalline displays
LIP	Light induced plating
LPCVD	Low pressure chemical vapor deposition
MFC	Mass flow controller
PC	Polycarbonate
PECVD	Plasma enhanced chemical vapor deposition
PLA	Polylactic acid
PV	Photovoltaic
RCA	Radio Corporation of America
SEM	Scanning electron microscope
SiO2:H	Hydrogenated silicon dioxide
TCO	Transparent conducting oxide
TE	Thermoelectric
TEOS	Trimethylchlorosilane
TFT	Thin-film transistor
USB	Universal-serial-bus
UV	Ultraviolet
XPS	X-ray photoelectron spectroscopy

List of characters

Symbol	Unit	Description
$\bar{x}$	–	The average rate
$\acute{E}$	$g \cdot m^{-2}s^{-1}$	Specific evaporation rate
A	mmHg	Antoine coefficient
A	Pa	Constant for calculating P_w and T_{dp} of water
AH	$mg \cdot m^{-3}$	Water vapor concentration or absolute humidity
AH_b	$g \cdot m^{-3}$	Vapor concentration into the water bubbler
AH_m	$g \cdot m^{-3}$	Vapor concentration into the humidity mixer
A_p	cm^2	Liquid surface area
A_s	cm^2	Surface area of the substrate
B	mmHg/K	Antoine coefficient
BG	eV	Band gap
C	K	Antoine coefficient
D	–	Constant
DS	V/cm	Dielectric strength
E	$g \cdot s^{-1}$	Evaporation rate
E	$g \cdot s^{-1}$	Evaporation rate
E_a	kJ/mol	Activation energy
ER	nm/s	Etch rate
E_{SiCl4}	$mg \cdot s^{-1}$ or $ml \cdot s^{-1}$	Evaporation rate from the $SiCl_4$ vaporizer
E_{SiCl4}	$mg \cdot s^{-1}$	Evaporation rate from the $SiCl_4$ evaporator
FF	-	Fill factor
j_{mpp}	mA/cm^2	Current at the maximum power point
j_{SC}	mA/cm^2	Short circuit current density
k	–	Extinction coefficient
$K_{c\,(SiCl4)}$	$mg \cdot m^{-3}$	mass concentration of $SiCl_4$ molecules into the gas stream

Symbol	Unit	Description
K_m	$m \cdot s^{-1}$	Mass transfer coefficient
M	$g \cdot mol^{-1}$	Molar mass (molar weight)
m	–	Constant for calculating P_w and T_{dp} of water
m_1 & m_2	mg	Reactant weights measured before and after evaporation
m_{H2O}	g	Weight of the water vapor held in the volume of dry gas
M_{H2O}	$g \cdot mole^{-1}$	Molar weight of H_2O molecules
m_{pp}	mW/cm^2	Maximum power point
MR	–	$SiCl_4$-to-H_2O molar ratio
MW	$g \cdot kmol^{-1}$	Molecular weight
N	–	Number of measured rates
n	–	Refractive index
N_{H2O}	mol	Mole number of H_2O molecules
N_{SiCl4}	mol	Mole number of $SiCl_4$ molecules
p	mW/cm^2	Power density
PFF	-	Pseudo fill factor
P_w	Pa	Partial water vapor pressure
P_{ws}	Pa	Saturation vapor pressure of water
Pv	mmHg	Saturated vapor pressure
$Q_{(b+m)}$	$l \cdot min^{-1}$	Total flow rate at the humidity mixer outlet
Q_b	$l \cdot min^{-1}$	Flow rate of N_2 gas into the water bubbler
Q_m	$l \cdot min^{-1}$	Flow rate of dry N_2 gas into the humidity mixer
Q_{SiCl4}	l/min	Flow rate of dry N_2 gas into the $SiCl_4$ evaporator
R	$J \cdot mol^{-1}K^{-1}$	The universal gas constant
RH	%	Relative humidity
RH_b	%	Alternative humidity at the outlet of the water bubbler
RH_m	%	Alternative humidity at the outlet of the humidity mixer
R_P	$\Omega\ cm^2$	Parallel resistance of the solar cell
R_S	$\Omega\ cm^2$	Series resistance of the solar cell

Symbol	Unit	Description
T	K	Temperature
T_a	°C	Annealing temperature
T_{APCVD}	°C	APCVD system temperature
T_b	°C	Boiling point
T_b	°C	Water bubbler-temperature
T_c	°C	$SiCl_4$ evaporator-temperature
TC	W/cm·K	Thermal conductivity
T_{dp}	°C	Dewpoint temperature
T_i	°C	Temperature of the liquid-gas interface
T_m	°C	Humidity mixer-temperature
T_n	°C	Constant for calculating P_w and T_{dp} of water
t_{SiO2}	nm	Thickness of the SiO_2 film
U_b	$mg \cdot s^{-1}$	Mass flow of the H_2O vapor through the bubbler
U_m	$mg \cdot s^{-1}$	Mass flow of the H_2O vapor through the mixer
v_{cg}	$cm \cdot s^{-1}$	Carrier-gas speed above the liquid surface
V_{mpp}	mV	Voltage at the maximum power pint
V_{OC}	mV	Open circuit voltage
WR	–	$SiCl_4$-to-H_2O mass ratio
M_{SiCl4}	$g \cdot mole^{-1}$	Molar weight of $SiCl_4$ molecules
x_n	–	Measured evaporation rate
ΔH	kJ/mol	Total enthalpy change of the chemical reaction
ΔH_{H2O}	kJ/mol	Enthalpy change of the O-H bond breaking
ΔH_{HCl}	kJ/mol	Enthalpy changes of the H-Cl bond coordinating
ΔH_{SiCl4}	kJ/mol	Enthalpy change of the Si-Cl bond breaking
ΔH_{SiO2}	kJ/mol	Enthalpy changes of the Si-O bond coordinating
ΔH_v	$kJ.mol^{-1}$	Heat of vaporization
Δn	–	Difference in refractive index before and after annealing
Δt_E	s	Measured evaporation period

Symbol	Unit	Description
Δt_{SiO2-x}	nm	Reduction in the SiO_{2-x} film thickness
$\Delta v_{n(SiCl4)}$	ml	Difference in the $SiCl_4$ volume within the vaporizer
η	%	Energy conversion efficiency
v_1 & v_2	ml	Reactant volumes measured before and after evaporation
v_{mix}	m^3	Volume of the mixture of water vapor and its holding gas
ρ	$g \cdot cm^{-3}$	Density
ρ_{SiCl4}	$g \cdot ml^{-1}$	Density of the evaporated $SiCl_4$
ρ_{SiO2}	$g \cdot cm^{-3}$	Measured density of the deposited SiO_2 film
σ	–	Standard deviation
Φ	W/m^2	Radiation power density

Frequently performed investigations

Investigation/experiments	Main contributor	Contributing colleagues
APCVD experiments* (**3600 ones**)	Esmail Issa	Dr. Henning Nagel
Maintenance of APCVD set up	Esmail Issa	--
CAD CAM design (30 ones)	Esmail Issa	--
ANSYS simulation (30 ones)	Esmail Issa	--
3D printing of CVD & LIP elements	Esmail Issa	Dr. Henning Nagel
Maintenance of 3D printer	Esmail Issa	Dr. Henning Nagel
FTIR measurements	Esmail Issa	--
FTIR measurement analysis	Esmail Issa	--
Ellipsometry measurements	Esmail Issa	Dr. Henning Nagel
Ellipsometry result analysis	Esmail Issa	Dr. Henning Nagel
Confocal laser microscopy	Esmail Issa	Hamed Javanbakht
Screen printing of polymer inks	Esmail Issa	Hamed Javanbakht & Eddy Coron & Lukasz Guzik
Electroplating on SHJ cells	Esmail Issa	Hamed Javanbakht & Eddy Coron
Ion milling of metallized samples	Esmail Issa	Aleksander Filipovic
Electroplating on PERC cells	Gisela Cimiotti	Gabriele Mikolasch & Jonas Eckert & Esmail Issa & Henning Nagel
Screen printing of Ag inks	Hamed Javanbakht	Lukasz Guzik

Investigation/experiments	Main contributor	Contributing colleagues
SEM/EDX inspections	Jutta Zielonka	Volker Kübler
I-V curve and cell performance	Elisabeth Schäffer	Felix Martin
HF dip of Si and PERC samples	Gabriele Mikolasch	G. Cimiotti & J. Eckert & H. Nagel
Laser ablation for PERC	Varun Arya	Stefan Schellinger
Resources & Lab support	Dr. Jonas Bartsch	Katharina Gensowski & Simon Rudolf
RCA Cleaned FZ Si wafers	Felix Schätzle	Clean room people
HNF cleaned FZ and CZ Si	Katrin Krieg	Felix Schätzle

* The APCVD experiments include SiO_{2-x} coatings on all wafers/cell-precursors, calcination and other treatment tests, and developing the system elements based on the deposition results.

List of publications

Esmail Issa, Henning Nagel, Jonas Bartsch, Markus Glatthaar and Edda Rädlein, Application of Hydrosilane-Free Atmospheric Pressure Chemical Vapor Deposition of SiO_x Films in The Manufacture of Crystalline Silicon Solar Cells. Thin Solid Films, 713, (2020)138338.

Esmail Issa, Henning Nagel, Lukasz Guzik, Hamed Javanbakht, Eddy Coron, Edda Rädlein and Stefan W. Glunz, Plated Front Side Metallization on Transparent Conducting Oxide Utilizing Low-Cost APCVD SiO_2 Insulating Layer, 36th EUPVSEC, Marseille, France. (2019) 431–435.

Henning Nagel, **Esmail Issa**, Timo Nagel, Markus Glatthaar and Stefan W. Glunz, Hydrosilane-Free Low-Cost APCVD of SiO_2 Films for Crystalline Si Solar Cell Applications, 36th EUPVSEC, Marseille, France. (2019) 419 – 422.

Patent applications

Granted patent

(30734-P-DE), Verfahren und Vorrichtung zum Drucken einer Struktur aus Druckmedium auf eine Oberfläche eines Substrats, Lukasz Guzik, Henning Nagel, **Esmail Issa**, 28.10.2019

Internal applications

ISE2 – 2019E62549 (I10449), Atmospheric pressure CVD method of chemically electrically and optically characteristic SiO_x/SiO_2 layers by the hydrolysis of $SiCl_4$, **Esmail Issa**, Henning Nagel, Markus Glatthaar. Received at the patent department of Fraunhofer Gesellschaft, Munich, 16.09.2019.

ISE2 – 2019E62549 (I10469), Conversion of the atmospheric pressure CVD kinetics from homogeneous to heterogeneous driven reaction for the formation of SiO_2 films, **Esmail Issa**, Henning Nagel. Received at the patent department of Fraunhofer Gesellschaft, Munich, 16.09.2019.

ISE2 – 2019E62549 (I10472), Design of an atmospheric pressure CVD for SiO_x/TiO_x films, **Esmail Issa**, Henning Nagel. Received at the patent department of Fraunhofer Gesellschaft, Munich, 16.09.2019.

Oral presentations

Esmail Issa, Henning Nagel, Edda Rädlein, Interaction of glassy thin layers of atmospheric pressure chemical vapor deposited SiO_x with Si wafers/solar cells, Living Glass Surface Conference X, Ilmenau, Germany, (2019).

Esmail Issa, Hennig Nagel, Edda Rädlein, Thin APCVD SiO_2 films for crystalline silicon solar cell applications, Living Glass Surface Conference IX, Ilmenau, Germany, (2017).

Esmail Issa, Hennig Nagel, Edda Rädlein, Design of an APCVD injector for the deposition of SiO_xN_y capping layers for Si solar cell applications, Exposure of Ph.D. Topic at the TU-Ilmenau Colloquium, Ilmenau, Germany, (2016).

Visual presentations

Esmail Issa, Henning Nagel, Lukasz Guzik, Hamed Javanbakht, Eddy Coron, Edda Rädlein and Stefan W. Glunz, Plated Front Side Metallization on Transparent Conducting Oxide Utilizing Low-Cost APCVD SiO_2 Insulating Layer, 36th EUPVSEC, Marseille, France. (2019).

Henning Nagel, **Esmail Issa**, Timo Nagel, Markus Glatthaar and Stefan W. Glunz, Hydrosilane-Free Low-Cost APCVD of SiO_2 Films for Crystalline Si Solar Cell Applications, 36th EUPVSEC, Marseille, France (2019).

Seminars

Esmail Issa, ANSYS simulation software for CFD, Doctoral Forum at the Fraunhofer institute for solar energy systems, Freiburg, Germany, (2019).

Esmail Issa, Introduction to the FTIR characterization, Doctoral Forum at the Fraunhofer institute for solar energy systems, Freiburg, Germany, (2017).

Declaration of contribution statement to parts of this research

Name	Contribution statement
Prof. Edda Rädlein*	Supervision & Methodology & Comprehensive validation and verification of the experiments and results & Reviewing and editing of publications & Visualization and colloquiums & Funding acquisitions & patent application approval & Communication support & Organizational support.
Dr. Henning Nagel	Choice of the research topic & Daily supervision & Scientific discussions & Experiment and result validation & Conceptualization & Methodology & Help with APCVD investigations & Laboratory resources & Help with PERC cell electroplating and HF dip & C^{++} programming/code writing and microcontroller installation & 3D printing and maintenance & Writing, reviewing and editing of publications & Patent application approval & Multiple characterizations & Visualizations and presentations & Contribution to the APCVD safety assessment documents & Introductions to related experiments/operations.
Dr. Jonas Bartsch	Team leading and organization & Work validation and result verification & Resources and laboratory support & Review and editing of publications & Visualization and presentation support & Doctoral seminars and scientific discussions & Additional topic methodologies.
Dr. Markus Glatthaar	Project administration & Result validation & Review and editing of publications & Visualization & Resources & Scientific discussions.
Dr. Sven Kluska	Team organization & PERC-solar cell electroplating investigations & Resources and laboratory support & Visualization and presentation support & Doctoral seminars & Introduction to the LEXT microscope.

Name	Contribution statement
Lukasz Guzik	Support with building the APCVD chambers & Performing of APCVD related safety assessment & Contribution to screen-printer self-manufacturing & PERC solar cell rear side Ag metallization & Support with screen-printing of polymer/Ag inks & Introduction to screen printing.
Aleksander Filipovic	Numerous safety and laboratory instructions & Resources and laboratory support & Introduction to the automated dicing saw & Introduction to the AST SHS-10RTA furnace & Introduction to the Hitachi ion milling machine & Support with ion milling.
Hamed Javanbakht (as a master student in the team)	Screen printing of Ag grid on SHJ cells & Contribution to 1) polymer screen printing of polymer inks, 2) electroplating of SHJ solar cells, 3) LEXT microscopy inspection, 4) and sample dicing and cleaning.
Eddy Coron (as a master student in the team)	Contribution to screen printing of polymer inks and electroplating on SHJ solar cells & Development of suited polymer inks.
Ahmad Ziade (as an assistant researcher)	Help with building the APCVD chambers and first injector installation.
Jutta Zielonka	Hundreds of SEM/EDX measurements & FIB investigations & Ion milling of Si coated Ge samples.
Volker Kübler	SEM/EDX measurements & Introduction to the Au sputtering machine & Laboratory safety instruction.
Gisela Cimiotti	Contribution to PERC electroplating & Introduction to the metallization laboratories, electroplating process, and cup-cell plater & HF dip of PERC cells & Introduction to XRF instrument & Support with LEXT inspection.
Gabriele Mikolasch	HF dip of Si and PERC samples & Contribution to the electroplating on PERC solar cells.
Jonas Eckert	Contribution to the HF dip and electroplating on PERC solar cells
Elisabeth Schäffer	I-V curve and solar cells' performance & Reflectance/Transmittance measurements on SiO_{2-x} coated Si samples (not presented in this book).

Name	Contribution statement
Felix Martin	I-V curve and solar cells' performance & Reflectance/Transmittance measurements on SiO_{2-x} coated Si samples (not presented in this book).
Katharina Gensowski	Resources & Safety instructions & Laboratory organization.
Simon Rudolf	Resources & Safety instructions & Laboratory organization.
Felix Schätzle	Supply with FZ Si and RCA cleaning.
Katrin Krieg	Supply with FZ and CZ Si and HNF cleaning.
Varun Arya	Laser ablation of PERC solar cell precursors.
Stefan Schellinger	Laser ablation of PERC solar cell precursors.
Leonard Krauss	Introduction to the Ellipsometer (Woollam-2000) and measurement support.
Dr. Frank Feldmann	Introduction to the Ellipsometer (Woollam-1600).
Dr. Charlotte Weiß	Introduction to the FTIR/ATR tool.
Dr. Stephanie Reiß*	XPS measurements and analysis on SiO_{2-x} coated Ge (TU-Ilmenau).
Dr. Kerstin Pfeifer*	SEM measurements of porous SiO_x-coated Si wafers (TU-Ilmenau).
Laura Stevens	AFM measurements and analysis of SiO_{2-x} coated Si wafers (not presented in this book).
Anand Kakadiya	Help with ANSYS simulation software.
Mechanical workshop people at ISE	Manufacturing of parts for the APCVD tool (financed by Fraunhofer ISE)

*Technical University of Ilmenau.

The other contributors are affiliated with Fraunhofer ISE.